Contemporary's

NUMBER POWER

Review

Whole Numbers to Algebra

ROBERT MITCHELL

Project Editor
Kathy Osmus

CB

CONTEMPORARY
BOOKS

CHICAGO

Library of Congress Cataloging-in-Publication Data

Mitchell, Robert, 1944–
    Number power review / Robert Mitchell.
        p.   cm.
    ISBN 0-8092-3805-5 (pbk.)
    1. Mathematics.   I. Title.
QA39.2.M575   1993
513'.142—dc20   93-8627
                                       CIP

Copyright © 1993 by Robert Mitchell
All rights reserved

No part of this publication may be reproduced, stored in
a retrieval system, or transmitted in any form or by any
means without the prior written permission of the
publisher.

Published by Contemporary Books, Inc.
Two Prudential Plaza, Chicago, Illinois 60601-6790
Manufactured in the United States of America
International Standard Book Number: 0-8092-3805-5

10   9   8   7   6   5

Published simultaneously in Canada by
Fitzhenry & Whiteside
195 Allstate Parkway
Markham, Ontario L3R 4T8
Canada

Editorial Director
Mark Boone

Editorial
Christine M. Benton
Katherine Willhoite
Eunice Hoshizaki
Leah Mayes
Lisa Black

Editorial Production Manager
Norma Fioretti

Production Editor
Jean Farley Brown

Illustrations
Graziano, Krafft & Zale, Inc.

Interior Design
Ophelia Chambliss-Jones

Typography
Impressions
Madison, Wisconsin

Contents

To the Instructor

Number Power Review is designed to help students brush up on basic computation and problem-solving skills—from whole numbers to beginning algebra. This book presents the core of mathematical skills considered most important in today's changing world. These are also the skills most likely to appear on state high school competency tests, on adult high school program tests, on the GED Test, and on pre-employment tests. Because of the importance of problem solving in all areas of math, *Number Power Review* discusses fifteen problem-solving strategies. These strategies are designed to help build the critical thinking skills students need today.

As math educators, we must recognize that today most computation performed outside the classroom is done either by estimating or by using a calculator. In response, *Number Power Review* integrates instruction in estimation and calculator use into all student activities. Moreover, students learn to identify problems for which estimating is a reasonable alternative to computing an exact answer. Students also learn that estimating can serve to check both computation accuracy and proper calculator use. Interestingly, research shows that students who develop proficiency in estimation and calculator use tend simultaneously to become better problem solvers.

Number Power Review is designed so that you, the instructor, can emphasize computation, estimation, and calculator use at your own level of comfort and at each student's level of ability. Whenever possible, each student should learn all three of these important skills. Ideally, students will practice estimation in designated exercises and strengthen calculator skills while checking computation answers.

To the Student

Welcome to *Number Power Review*. This book is designed to strengthen your basic math and problem-solving skills to help prepare you for various basic skills tests. Included are state high school competency tests, adult high school completion tests, the GED Test, and pre-employment tests.

Special features of *Number Power Review* include
- nine study units covering core math skills
- fifteen problem-solving strategies designed to give you confidence with word problems
- instruction in estimating and in using a calculator
- "Test Readiness Checkup" exercises at the end of Units 2–9 to allow you to apply your skills in a test-like format
- a practice test to help identify those skills you may wish to review and to check your readiness to take other basic skills math tests

To get the most out of your work, do each problem carefully. Check your answers to make sure you are working accurately. A complete answer key starts on page 238.

UNIT 1 Number Sense

Writing Familiar Numbers

Most numbers you'll ever use will have four or fewer digits.
1,326 is an example of a four-digit number.
   Notice how the **value** of each digit is determined by its **place** in the number.

The First Four Place Values

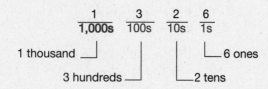

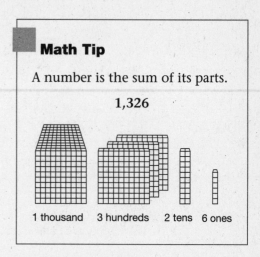

Math Tip

A number is the sum of its parts.

1,326

1 thousand   3 hundreds   2 tens   6 ones

Zero is often used as a **placeholder** to show that there are no ones, tens, or hundreds.

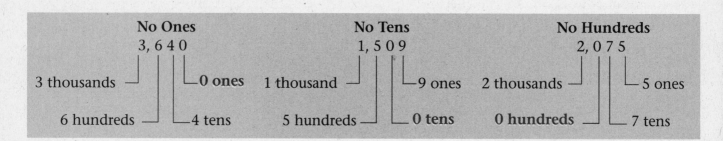

Look at this example and the rules that follow.
1,973 is read "one thousand, nine hundred seventy-three."
• Do not use the word *and* when writing or saying whole numbers.
• Write a hyphen (-) in compound words like twenty-one (21) and ninety-nine (99).
• Place a comma after the word *thousand,* but not after the word *hundred.*

Practice

▮ **Write each number below in words.**

1. 46 _____

2. 381 _____

3. 1,508 _____

4. 7,062 _____

Writing Larger Numbers

In larger numbers, commas separate digits into groups of three. Notice each pattern of 100s, 10s, and 1s.

Millions	Thousands	Ones

___ ___ ___ , ___ ___ ___ , ___ ___ ___
100s 10s 1s   100s 10s 1s   100s 10s 1s

To read a larger number, read each group of digits separately.

239,710,579 is read 239 *million*, 710 *thousand*, 579.

↑ millions
↑ thousands
↑ ones

Note: Commas are used to separate millions from thousands and thousands from ones.

Math Tip

To strengthen skills for reading larger numbers, memorize these:

Number	Read as
1,000	1 thousand
10,000	10 thousand
100,000	100 thousand
1,000,000	1 million
10,000,000	10 million
100,000,000	100 million

Practice

Match each number with its word meaning.

Number	Word Meaning
____ 1. 306,000	a) three million, six thousand
____ 2. 3,600,000	b) thirty-six million
____ 3. 36,000,000	c) three million, six hundred thousand
____ 4. 3,006,000	d) thirty million, sixty thousand
____ 5. 360,000	e) three hundred six thousand
____ 6. 30,060,000	f) three hundred sixty thousand

Write each number using digits.

7. sixty-one thousand, eight hundred seventy-two _____

8. four hundred sixty-two thousand, eight hundred _____

9. five million, eight hundred thirty-five thousand _____

Write the following numbers in words.

10. diameter of the earth: 7,926 miles _____

11. distance of the moon from the earth: 237,300 miles _____

12. distance of the sun from the earth: 92,900,000 miles _____

Comparing and Ordering Numbers

Symbols are used to show the relationship between two numbers.

Symbol	Meaning	Example	Read as
>	is greater than	9 > 5	"9 is greater than 5"
<	is less than	3 < 11	"3 is less than 11"
=	is equal to	6 = 6	"6 is equal to 6"
≠	is not equal to	8 ≠ 12	"8 is not equal to 12"

To compare two numbers, compare digits from left to right. Be sure to compare digits that have the same place value.

Compare 126 and 98.
1|2 6 Because 1 > 0,
 9 8 126 > 98.

Compare 198 and 249.
1|9 8 Because 1 < 2,
2|4 9 198 < 249.

Compare 361 and 347.
3|6|1 Since 3 = 3, compare 6 to 4: 6 > 4,
3|4|7 so 361 > 347.

Practice

Part A. Write >, < , or = to compare each pair of numbers below.

1. 38 _____ 42   30 _____ 30 $129 _____ $124

2. 819 _____ 822   1,035 _____ 872 $2,487 _____ $2,574

3. 7,975 _____ 7,975 $5,052 _____ $5,061   25,435 _____ 25,431

4. $6.38 _____ $6.78 $14.61 _____ $13.73 $25.18 _____ $34.06

Part B. Look at the Employee Loan Amounts list shown below. Rewrite the list in order of loan amount. List the largest amount first (1st) and so on.

Loan Amount	Loan Amount	Employee Name	Loan Amount
1st: _____	4th: _____	Cheung	$4,867
		Cooke	$5,932
2nd: _____	5th: _____	Ferguson	$6,779
		Herzberg	$5,589
3rd: _____	6th: _____	Nason	$6,780
		Ramirez	$4,814

Understanding Number Sentences

A **number sentence** is a mathematical statement that compares numbers or **numerical expressions**—numbers joined by +, −, ×, and ÷ signs. A number sentence may be an **equation** or an **inequality**.

Math Tip

Test questions often ask you to

- decide if a number sentence is true or false
- complete a number sentence by finding the number that makes it true

Equation   **Inequalities**

$24 = 14 + 10$ $106 > 80 + 9$ $63 < 90 - 20$ $14 \neq 7 + 9$

- An **equation** contains the **equals sign** (=).
- An **inequality** contains an **inequality sign** (>, <, or ≠).

To see if a number sentence is true, perform all operations (+, −, ×, or ÷) and then compare the resulting two numbers.

Which of the following number sentences are true?
a) $23 + 10 > 35$ **False**, because $23 + 10 = 33$, and $33 < 35$.
b) $34 - 12 = 20$ **False**, because $34 - 12 = 22$, *not* 20.
c) $30 \div 6 < 6$ **True**, because $30 \div 6 = 5$, and $5 < 6$.
d) $14 \times 3 \neq 51$ **True**, because $14 \times 3 = 42$, and $42 \neq 51$.

Practice

In each number sentence below, perform the indicated operation. Then circle **T** if the sentence is true or **F** if it is false.

1. $9 < 5 + 7$ T or F $14 - 9 > 3$ T or F $13 + 12 \neq 25$ T or F

2. $11 > 17 - 6$ T or F $15 - 7 = 9$ T or F $3 \times 4 > 14$ T or F

3. $20 \neq 2 \times 9$ T or F $9 < 56 \div 7$ T or F $42 \div 6 > 7$ T or F

4. $5 \times 8 > 38$ T or F $36 \div 4 < 8$ T or F $65 > 9 \times 7$ T or F

Write a *whole number* to make each equation or inequality true. For each inequality, more than one answer is correct.

5. $5 + \underline{\quad} = 12$ $7 \times \underline{\quad} = 42$ $9 + 8 \neq 12 + \underline{\quad}$

6. $45 \div 9 > \underline{\quad}$ $30 - \underline{\quad} < 5 \times 2$ $4 \times 8 < \underline{\quad}$

7. $14 - 8 > \underline{\quad}$ $3 \times 4 > \underline{\quad}$ $25 \div \underline{\quad} > 5$

8. $9 \times \underline{\quad} = 54$ $8 - \underline{\quad} > 3$ $32 \div \underline{\quad} = 4$

Rounding Whole Numbers

Often, it is useful to **estimate**—to give a number that is "about equal" to an exact amount. One way to estimate is to use round numbers.

The newspaper says that 800 people attended the picnic. Actually, 829 people were there. The number 800 is called a **round number**. To the hundreds place, 829 **rounds to** 800. To the thousands place, 829 rounds to 1,000.

> **Math Tip**
>
> A round number usually ends in one or more zeros.

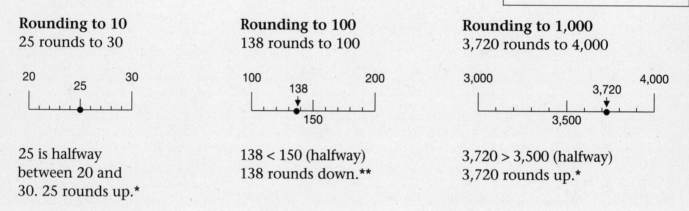

Rounding to 10
25 rounds to 30

25 is halfway between 20 and 30. 25 rounds up.*

Rounding to 100
138 rounds to 100

138 < 150 (halfway)
138 rounds down.**

Rounding to 1,000
3,720 rounds to 4,000

3,720 > 3,500 (halfway)
3,720 rounds up.*

* A number that's *halfway or more* to the larger round number always rounds *up* to the larger number.
** A number that's *less than halfway* to the larger round number always rounds *down* to the smaller number.

Practice

▨ Round each number. Circle each answer choice.

To the nearest ten	To the nearest hundred	To the nearest thousand
1. 42: 40 or 50	175: 100 or 200	1,198: 1,000 or 2,000
2. 67: 60 or 70	750: 700 or 800	4,723: 4,000 or 5,000

▨ Round each number to the nearest ten.

3. 35 _____   73 _____   19 _____   84 _____   95 _____

▨ Round each number to the nearest hundred.

4. 109 _____   350 _____   284 _____   420 _____   834 _____

▨ Round each number to the nearest thousand.

5. 3,058 _____   1,942 _____   7,498 _____   4,500 _____   6,056 _____

Rounding Dollars and Cents

You most often round money to the nearest $1.00 or $10.00.

Rounding to the Nearest Dollar
Look at the number of cents.

$0.<u>8</u>3 rounds to $1.00
↑
Round up if
50¢ or more.

$3.<u>2</u>9 rounds to $3.00
↑
Round down if
less than 50¢.

$12.<u>5</u>0 rounds to $13.00
↑
Round up if
50¢ or more.

Rounding to the Nearest Ten Dollars
Look at the number of dollars.

$<u>9</u>.18 rounds to $10.00
↑
Round up if $5 or more.

$4<u>1</u>.84 rounds to $40.00
↑
Round down if less than $5.

$17<u>5</u>.00 rounds to $180.00
↑
Round up if $5 or more.

Practice

■ **Round each purchase to the nearest dollar.**

1.

$1.89

$4.28

$11.50

$13.25

■ **Round each purchase to the nearest ten dollars.**

2.

$19.75

$32.27

$107.50

$88.88

■ **Round these larger amounts.**

3. Clay's monthly rent of $529

_____
(nearest $100)

5. the cost of a $3,819 motorcycle

_____
(nearest $1,000)

4. Enrico's monthly salary of $1,675

_____
(nearest $100)

6. the cost of Henry's new $14,395 car

_____
(nearest $1,000)

Estimating Sums

Estimating, your most important math tool, can be used to find an approximate answer (estimate). You estimate to
- find a "close" answer when that's all you need
- check calculation accuracy
- check calculator answers
- help you choose answers on tests

Math Tip

When estimating, no single answer is the only correct one. Any reasonable estimate is fine. The examples below show one way to estimate: by rounding each number to the same place.

To estimate a sum, round each number. Then add the round numbers.

Estimate: 119 + 72
Round each number to the tens place. Add.

$$119 \rightarrow 120$$
$$+ 72 \rightarrow + 70$$
$$\overline{190}$$

Estimate: 479 + 318
Round each number to the hundreds place. Add.

$$479 \rightarrow 500$$
$$+ 318 \rightarrow + 300$$
$$\overline{800}$$

Estimate: 1,191 + 594
Round each number to the hundreds place. Add.

$$1,191 \rightarrow 1,200$$
$$+ 594 \rightarrow + 600$$
$$\overline{1,800}$$

Practice

Estimate these sums. Any reasonable estimate is acceptable.

1.
$$\begin{array}{r} 56 \\ + 29 \\ \hline \end{array}$$
$$\begin{array}{r} 73 \\ + 52 \\ \hline \end{array}$$
$$\begin{array}{r} 327 \\ + 188 \\ \hline \end{array}$$
$$\begin{array}{r} 493 \\ + 218 \\ \hline \end{array}$$

2.
$$\begin{array}{r} 1,193 \\ + 720 \\ \hline \end{array}$$
$$\begin{array}{r} 3,265 \\ + 514 \\ \hline \end{array}$$
$$\begin{array}{r} 3,274 \\ + 1,865 \\ \hline \end{array}$$
$$\begin{array}{r} 5,878 \\ + 2,387 \\ \hline \end{array}$$

3.
$$\begin{array}{r} \$7.89 \\ + 5.14 \\ \hline \end{array}$$
$$\begin{array}{r} \$13.45 \\ + 4.89 \\ \hline \end{array}$$
$$\begin{array}{r} \$125.99 \\ + 62.84 \\ \hline \end{array}$$
$$\begin{array}{r} \$119.23 \\ + 71.84 \\ \hline \end{array}$$

Estimate an answer to each problem below. Then, using only your estimate, choose the exact answer.

4.
$$\begin{array}{r} 391 \\ 207 \\ + 193 \\ \hline \end{array}$$
  (1) 641
  (2) 791
  (3) 921

5.
$$\begin{array}{r} \$19.59 \\ 14.98 \\ + 10.25 \\ \hline \end{array}$$
  (1) \$34.92
  (2) \$40.32
  (3) \$44.82

6.
$$\begin{array}{r} 2,129 \\ 1,082 \\ + 539 \\ \hline \end{array}$$
  (1) 3,750
  (2) 4,470
  (3) 5,290

Estimating Differences

To estimate a difference, round each number to the same place. Then subtract the round numbers.

Math Tip

Estimating works well on multiple-choice questions when answer choices are not too close together in value. (See problems 4–6 on page 8.)

Estimating helps here   **but not here!**

Subtract: Subtract:

	Estimate				Estimate	
301	300	(1) 81		407	400	(1) 208
− 196	− 200	*(2) 105		− 189	− 200	(2) 213
	100	(3) 123			200	*(3) 218

Estimating works well   Answer choices are too
because answer choices   close in value for the
are not close in value.   estimate to help.

*correct answer

Practice

Estimate these differences. Any reasonable estimate is acceptable.

1.	37	172	67	442
	− 19	− 99	− 26	− 116

2.	1,314	1,168	3,289	5,110
	− 829	− 593	− 1,317	− 2,967

3.	$8.93	$17.18	$113.12	$142.50
	− 4.16	− 9.24	− 71.92	− 98.78

Use only estimation to help you choose each exact answer below. Circle the problem for which estimating doesn't help.

4.	686	(1) 268	5.	$48.64	(1) $28.66	6.	3,963	(1) 1,294
	− 298	(2) 388		− 19.98	(2) $29.06		− 1,979	(2) 1,674
		(3) 528			(3) $29.76			(3) 1,984

Estimating Products

One way to estimate a product is to round each number to its highest place; each number will then contain a single digit and one or more zeros. Then multiply the round numbers.

Math Tip

To multiply numbers that end in zero

- multiply the nonzero digits
- add on the number of zeros in the two numbers

$$\begin{array}{rl} 300 & \text{2 zeros} \\ \times\ 20 & +\ \text{1 zero} \\ \hline 6,000 \leftarrow & \text{3 zeros} \\ \uparrow & \\ 3 \times 2 & \end{array}$$

Estimate: 94×7
Round 94 and multiply.

$$\begin{array}{rcr} 94 & \to & 90 \\ \times\ 7 & \to & \times\ 7 \\ \hline & & 630 \end{array}$$

Estimate: $\$79 \times 21$
Round each number to the nearest 10. Multiply.

$$\begin{array}{rcr} \$79 & \to & \$80 \\ \times\ 21 & \to & \times\ 20 \\ \hline & & \$1,600 \end{array}$$

Estimate: 213×187
Round each number to the nearest 100. Multiply.

$$\begin{array}{rcr} 213 & \to & 200 \\ \times\ 187 & \to & \times\ 200 \\ \hline & & 40,000 \end{array}$$

Practice

Estimate these products. Any reasonable estimate is acceptable.

1. $\begin{array}{r} 37 \\ \times\ 8 \\ \hline \end{array}$
    $\begin{array}{r} 81 \\ \times\ 9 \\ \hline \end{array}$
    $\begin{array}{r} 41 \\ \times\ 32 \\ \hline \end{array}$
    $\begin{array}{r} 58 \\ \times\ 19 \\ \hline \end{array}$

2. $\begin{array}{r} 112 \\ \times\ 23 \\ \hline \end{array}$
    $\begin{array}{r} 273 \\ \times\ 49 \\ \hline \end{array}$
    $\begin{array}{r} 302 \\ \times\ 116 \\ \hline \end{array}$
    $\begin{array}{r} 593 \\ \times\ 287 \\ \hline \end{array}$

3. $\begin{array}{r} \$7.89 \\ \times\ 8 \\ \hline \end{array}$
    $\begin{array}{r} \$9.31 \\ \times\ 6 \\ \hline \end{array}$
    $\begin{array}{r} \$56.99 \\ \times\ 11 \\ \hline \end{array}$
    $\begin{array}{r} \$89.75 \\ \times\ 19 \\ \hline \end{array}$

In each problem below, choose the best estimate of the exact answer.

4. $\begin{array}{r} 79 \\ \times\ 61 \\ \hline \end{array}$
    (1) 100×50
    (2) 80×60
    (3) 80×50

5. $\begin{array}{r} 113 \\ \times\ 98 \\ \hline \end{array}$
    (1) 125×100
    (2) 100×100
    (3) 110×100

6. $\begin{array}{r} \$139.87 \\ \times\ 49 \\ \hline \end{array}$
    (1) $\$140 \times 50$
    (2) $\$150 \times 50$
    (3) $\$100 \times 50$

Using Clustering to Estimate a Large Sum

Multiplication can often be used to estimate the value of a large sum. When a group of numbers **clusters** around a common value, you estimate the total by *multiplying* the common value by the number of amounts in the group.

Estimate the sum.	1,198
	1,047
The numbers cluster	983
around 1,000.	+ 856
A good estimate is:	

Total = 1,000 × 4

= 4,000

Note: When possible, choose a round number ending in 0s as the *common value*.

Estimate the total number of houses sold during the months shown.

The numbers cluster around 300.
A good estimate is:

Total = 300 × 5

= 1,500 houses

Houses Sold	
Month	**Sales**
March	321
April	287
May	296
June	345
July	307
Est. Total:	

Practice

■ **For each problem, find the common value that the numbers cluster around. Then estimate a sum.**

1. 3,275   **a)** Common value: _____
    3,092
    2,947   **b)** Estimated sum: _____
    + 2,890

2. $2,147   **a)** Common value: _____
    1,908
    2,009   **b)** Estimated sum: _____
    +1,846

3. Estimate the total: $489.50 + $473.80 + $536.90 + $518.00 + $507.75.

   a) Common value: _____

   b) Estimated sum: _____

4. Estimate the total hamburger sales during the five weekdays.

Day	Sales
Monday	693
Tuesday	746
Wednesday	613
Thursday	694
Friday	715
Est. Total:	

5. Estimate the total number of miles driven by the four drivers listed.

Miles Driven	
Driver	**Miles**
Kushmir	1,983
Atwood	2,176
Burns	1,857
Rayas	2,209
Est. Total:	

Estimating Quotients

Dividing by a One-Digit Number

To estimate when dividing by a one-digit number, use compatible numbers. **Compatible numbers** are numbers that divide evenly—having no remainder. Compatible numbers come from basic multiplication facts:

$$\overset{7}{4\overline{)28}}$$ 4 and 28 are compatible numbers.

To estimate a quotient when you are dividing by a one-digit divisor
- round the first digit or two of the dividend to a number that is compatible with the divisor
- write 0s for the other digits of the dividend
- divide the compatible digits and write 0s as placeholders in the quotient

> **Math Tip**
>
> - The number being divided is called the *dividend*.
> - The number the dividend is divided by is called the *divisor*.
> - The answer is called the *quotient*.
>
> $$\overset{6 \leftarrow \text{quotient}}{7\overline{)4\,2}}$$
> divisor ↗   ↖ dividend

Problem: $7\overline{)37}$

Recall: $7 \times 5 = 35$
Round 37 to 35.

Divide: $\overset{5}{7\overline{)35}}$

Estimate: 5

Problem: $8\overline{)719}$

Recall: $8 \times 9 = 72$
Round 719 to 720.
Write a place-holding 0.
↓
Divide: $\overset{90}{8\overline{)720}}$

Estimate: 90

Problem: $6\overline{)5,146}$

Recall: $6 \times 8 = 48$
Round 5,146 to 4,800.
Write place-holding 0s.
↓↓
Divide: $\overset{800}{6\overline{)4,800}}$

Estimate: 800

Note: You can also round 5,146 to 5,400 and get an estimate of 900.

Practice

Estimate these quotients. Any reasonable estimate is acceptable.

1. $6\overline{)43}$ $5\overline{)29}$ $7\overline{)24}$ $9\overline{)\$38}$

2. $7\overline{)489}$ $8\overline{)\$251}$ $3\overline{)817}$ $4\overline{)\$315}$

3. $2\overline{)\$3{,}994}$ $5\overline{)1{,}439}$ $7\overline{)\$5{,}179}$ $8\overline{)7{,}357}$

Dividing by a Two-Digit Number

One way to estimate when the divisor contains more than 1 digit is to round both divisor and dividend.

- Round to numbers that make division easy for you. Usually, this means rounding to numbers where the first digits of the divisor and dividend are compatible numbers.
- Divide and write zeros as placeholders when needed.

Problem: $23\overline{)827}$	**Problem:** $89\overline{)4{,}609}$
Think: $8 \div 2 = 4$	Think: $45 \div 9 = 5$
Round 23 to 20 and 827 to 800.	Round 89 to 90 and 4,609 to 4,500.
Divide: $20\overline{)80\!\!\!/0\!\!\!/}$ with 40 above	Divide: $90\!\!\!/\overline{)4{,}50\!\!\!/0\!\!\!/}$ with 50 above
Step 1. Cross out one 0 in both divisor and dividend.	*Step 1.* Cross out one 0 in both divisor and dividend.
Step 2. Divide 80 by 2.	*Step 2.* Divide 450 by 9.
Estimate: 40	**Estimate: 50**
	Note: You can round 89 to 100 and 4,609 to 5,000 and still get the same estimate of 50. Or you can round 89 to 100 and 4,609 to 4,600 and get an estimate of 46. Either is fine.

Practice

■ **Estimate these quotients.**

4. $21\overline{)86}$ $32\overline{)\$90}$ $11\overline{)20}$ $19\overline{)\$80}$

5. $39\overline{)129}$ $51\overline{)488}$ $21\overline{)\$837}$ $17\overline{)527}$

6. $82\overline{)4{,}390}$ $28\overline{)\$7{,}148}$ $41\overline{)3{,}840}$ $72\overline{)\$12{,}343}$

Data Highlight: Reading a Table

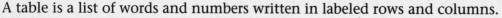

A table is a list of words and numbers written in labeled rows and columns.
- Rows are read across.   **Row:** 42 25 36 →

- Columns are read from top to bottom. **Column:**

28
19
35 ↓

Look at the nutrition table below.

Nutritional Information for Selected Fast-Food Sandwiches
(Typical values are given.)

Food Item	Calories	Protein (g*)	Fat (g*)
Hamburger	428	28	19
Cheeseburger	470	31	23
Fish Sandwich	344	33	10
Chicken Sandwich	378	32	12
Ham Sandwich	451	30	21

*g = grams   A gram is a metric weight unit.
A raisin weighs about one gram.

Example 1

How many grams of protein does
a chicken sandwich contain?

Find the intersection of the row
labeled Chicken Sandwich and the
column labeled Protein.

Protein
28 Read down.
31
33
Chicken Sandwich 378 32 ↓
Read across.

Answer: 32 g

Example 2

Which of the listed sandwiches
is lowest in fat?

Scan down the column labeled Fat and
choose the smallest number: 10. Read
the label of the row that contains the 10.

Fat
19 Read down.
23 ↓
Fish Sandwich 344 33 10
12
Look left for label.

Answer: Fish Sandwich

Practice

■ **Problems 1–5 refer to the nutrition table on page 14.**

1. How many calories does a cheeseburger contain?

2. How many grams of fat does a chicken sandwich contain?

3. According to the table, which sandwich has the:
   a) most calories? _____   c) most fat? _____
   b) least amount of protein? _____   d) fewest calories? _____

4. A sandwich with a high   (1) protein
   fat content tends to have   (2) weight
   a high amount of: (3) calories

5. A sandwich with a high   (1) fat
   calorie content tends to   (2) protein
   have a high amount of: (3) vitamins

■ **Problems 6–9 are based on the portion of the federal income tax table shown below.**

If Taxable income is—		And you are—			
At least	But less than	Single	Married filing jointly	Married filing separately	Head of a household
			Your Tax is—		
$20,800	$20,850	$3,186	$3,124	$3,621	$3,124
20,850	20,900	3,200	3,131	3,635	3,131
20,900	20,950	3,214	3,139	3,649	3,139
20,950	21,000	3,228	3,146	3,663	3,146
21,000	21,050	3,242	3,154	3,677	3,161

6. Ben is single and has a taxable income of $20,945. How much tax does he owe?

   Tax owed: _____

7. Anne is married and is filing jointly with her husband. If their taxable income is $21,030, how much tax do they owe?

   Tax owed: _____

8. Julia is "head of a household." If she owes $3,131 in taxes, her taxable income is between

   $ _____ and $ _____.

9. José is married but is filing separately from his wife. If José must pay $3,649 in taxes, his taxable income is between

   $ _____ and $ _____.

Number Highlight: **Negative Numbers**

Along with the **positive numbers** we've been studying in this chapter, there are also **negative numbers**.
- Positive numbers are numbers larger than zero.
- Negative numbers are numbers smaller than zero.

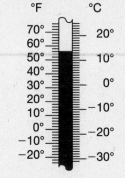

You may already know some of the common uses of negative numbers:
- reading temperatures below zero
- recording yardage losses in football
- reporting drops in stock market prices
- measuring distances below sea level

Always write a negative (minus) sign in front of a negative number:

Current Temperature   **Stock Market Changes**
    $-12°$ Celsius   IBM -3

A thermometer contains both *positive* (10°) and *negative* temperature readings ($-10°$).

The Number Line

A **number line** looks like a thermometer scale on its side. Negative numbers are written to the left of zero; positive numbers are written to the right.

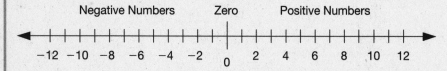

Numbers on a number line get larger as you move from left to right. Every negative number is smaller than 0. Every positive number is larger than 0.

Examples: $-8 < -6$ $-4 < 0$ $-2 > -5$ $6 > -8$ $-1 < 1$

Addition and subtraction of negative numbers are easily visualized on a number line. To *add*, start at the 1st number and move the 2nd number of spaces to the answer.
- If the 2nd number is positive, move to the right of the first number.
- If the 2nd number is negative, move to the left of the first number.

Add: $-3 + 8$ Add: $4 + (-6)$

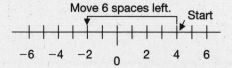

Answer: 5   Answer: -2

To *subtract,* simply count the number of spaces between the two numbers.
- The answer is positive when you subtract a smaller number from a larger number.
- The answer is negative when you subtract a larger number from a smaller number.

Subtract: 1 − (−3)   Subtract: 2 − 5

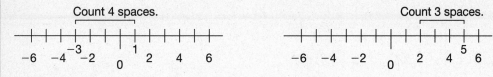

Answer: 4   Answer: −3

Practice

■ **Write >, <, or = to compare each pair of numbers below.**

1. −7 ____ −3   −6 ____ 0   −2 ____ 2   5 ____ −3

2. 0 ____ −1   3 ____ −2   −4 ____ 8   0 ____ 4

■ **Show how numbers are added and subtracted on each number line below.**

3. Add: 5 + (−3)   5. Subtract: 2 − (−4)

4. Add: 6 + (−2)   6. Subtract: −3 − (−2)

■ **Solve the problems. Sketch number lines if you find it helpful.**

7. At 2:00 A.M. Sunday, the temperature in Peoria was −5°F. By 6:00 A.M., the temperature had risen by 8°. What was the temperature in Peoria at 6:00 A.M. Sunday?

8. An airplane flying at 800 feet above sea level detects a submarine directly below, submerged at a depth of −200 feet (200 feet below sea level). How high is the plane above the submarine?

Calculator Spotlight

Learning to use a calculator quickly and confidently is an important math skill. With *Number Power Review*, you'll learn to use a calculator to solve a variety of math problems.

The Calculator Keyboard

The calculator below is similar to one you've seen or one you may be using.

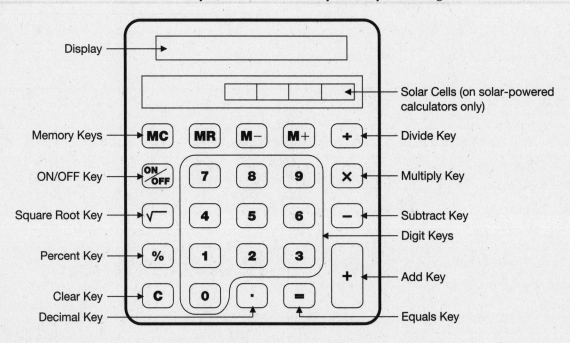

Locate the following keys (or similar keys) on your calculator.

- The **on/off key** (**ON/OFF**). You press (**ON/OFF**) once to turn a calculator on and press it again to turn it off. Some calculators have separate (**ON**) and (**OFF**) keys.

- The **digit keys** (**0**), (**1**), (**2**), (**3**), (**4**), (**5**), (**6**), (**7**), (**8**), and (**9**). To enter a number on a calculator, you simply press one digit at a time.

- The **decimal point key** (**·**) To enter cents, you press (**·**) to separate dollars from cents.

- The **function keys** (**+**), (**−**), (**×**), and (**÷**). You press a function key each time you add, subtract, multiply, or divide.

- The **clear key** (**C**). Pressing (**C**) erases the display. You press (**C**) each time you begin a new problem or when you've made a keying error.

Different calculators use different clear key symbols. Other commonly used symbols are:

(**ON/C**) On/Clear   (**CE/C**) Clear Entry/Clear   (**CE**) Clear Entry   (**AC**) All Clear

Displayed Numbers

Enter 4,730 on your calculator.

Press Keys   **Display Reads**

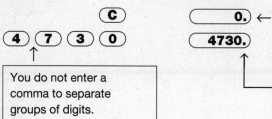

Most calculators display a **0.** when first turned on and after the clear key is pressed.	

You do not enter a comma to separate groups of digits.

Most calculators display a decimal point to the right of a whole number.

Enter $5.94 on your calculator.

Press Keys   **Display Reads**

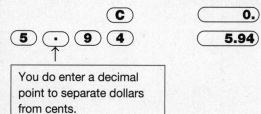

Calculator Keys

A calculator does not have a comma (,) key or a dollar sign ($) key.

You do enter a decimal point to separate dollars from cents.

Practice

Use a calculator to answer the following questions.

1. With your calculator on, press the whole-number keys:

   ① ② ③ ④ ⑤ ⑥ ⑦ ⑧ ⑨

   a) How many digits appear on this display? _____

   b) What is the largest number your calculator can display?

   _____

2. What key do you press to clear the display? _____

3. What appears on the display after you push the clear key? _____

4. Enter the following on your calculator. Then show how the calculator displays each number or amount.

Enter	**Display Reads**	**Enter**	**Display Reads**
a) $.58	_____	c) 239	_____
b) $8.45	_____	d) 1,956	_____

Number Sense Checkup

Circle each answer choice. Check your answers on page 239; then correct any errors.

1. How is 2,043 written in words?

   (1) two thousand, four hundred three
   (2) twenty thousand, forty-three
   (3) two thousand and forty-three
   (4) two thousand, four hundred thirty
   (5) two thousand forty-three

2. In 108,619, what does 0 stand for?

   (1) no tens
   (2) no hundreds
   (3) no thousands
   (4) no ten thousands
   (5) no hundred thousands

3. How is three million, five hundred thousand, four hundred sixty written in digits?

   (1) 3,500,046
   (2) 3,500,460
   (3) 3,504,600
   (4) 3,540,600
   (5) 3,546,000

4. Gina earns $6.45 per hour, and her friend Maria earns $7.12. Which expression comparing these two amounts is true?

   (1) $6.45 > $7.12
   (2) $6.45 = $7.12
   (3) $6.45 < $7.12
   (4) $7.12 < $6.45
   (5) $7.12 = $6.45

5. For which whole number is the following inequality true?
   $5 \times \underline{} > 34$

   (1) 3   (4) 6
   (2) 4   (5) 7
   (3) 5

6. The direct distance (by air) from New York to San Francisco is 3,267 miles. Round this distance to the nearest hundred miles.

   (1) 3,000
   (2) 3,100
   (3) 3,200
   (4) 3,300
   (5) 3,400

Problem 7 refers to the drawing.

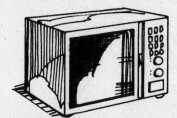

Better Homes
Microwave

$374.99

7. Rounded to the nearest $10, what is the price of the microwave oven?

   (1) $350
   (2) $360
   (3) $370
   (4) $375
   (5) $400

Problem 8 refers to the following list.

Day	Retail Sales
Monday	$379.85
Tuesday	$412.90
Wednesday	$406.25
Thursday	$387.40
Friday	$397.96
Total:	

8. Estimate the total retail sales for the week shown.

   (1) $1,600   (4) $2,200
   (2) $1,800   (5) $2,400
   (3) $2,000

9. Which of the following will give the best estimate of the purchase prices shown at right?

(1)	$5.00	(2)	$5.00	(3)	$6.00	(4)	$6.00	(5)	$6.00
	3.00		2.00		2.00		3.00		3.00
	+ 1.00		+ 1.00		+ 1.00		+ 1.00		+ 2.00

$6.39
$2.07
$.89

In problems 10–13, use only estimation to help you choose each exact answer.

10. $3,087 + 2,943 =$

    (1) 5,240   (4) 6,420
    (2) 5,660   (5) 6,870
    (3) 6,030

11. $\$823 - \$297 =$

    (1) $526   (4) $796
    (2) $606   (5) $846
    (3) $716

12. $\$98 \times 21 =$

    (1) 1,362   (4) 2,058
    (2) 1,572   (5) 2,398
    (3) 1,802

13. $19 \overline{)4,047}$

    (1) 27   (4) 175
    (2) 82   (5) 213
    (3) 138

14. The lowest recorded temperature in the U.S., −79°F, occurred on January 23, 1971, in Prospect Creek, Alaska. About how many degrees below freezing (32°F) did the temperature drop that day?

    (1) 30°F   (4) 80°F
    (2) 40°F   (5) 110°F
    (3) 70°F

Problem 15 refers to the partial tax table.

Amount of Sale	Tax
$18.75 to $18.91	$1.13
18.92 to 19.08	1.14
19.09 to 19.24	1.15
19.25 to 19.41	1.16
19.42 to 19.58	1.17

15. Using the partial sales tax table, determine the amount of tax on a $19.16 purchase.

    (1) $1.13   (4) $1.16
    (2) $1.14   (5) $1.17
    (3) $1.15

Problem 16 refers to the nutrition table.

Nutritional Values of Selected Grains		
(Each value is for a cooked serving of one cup.)		
Grain	Calories	Protein (grams)
A. White Rice	186	3.7
B. Brown Rice	232	4.9
C. Oats	132	4.8
D. Rye	237	8.6
E. Wheat	221	4.8

16. Which of the following lists the grains shown on the table in the order of calories—listing the grain with the most calories first and so on?

    (1) B, D, A, E, C
    (2) D, B, C, E, A
    (3) C, A, E, B, D
    (4) B, C, A, E, D
    (5) D, B, E, A, C

Adding Whole Numbers

Math Tip

When adding, **regrouping** can also be called *carrying* or *renaming*.

To add two or more numbers
- add from right to left, starting with the ones column
- regroup as needed

Use estimation or a calculator to check your answers.

1. Add the ones: 13. Write 3; regroup 10 ones as 1 ten.

2. Add the tens: 17. Write 7; regroup 10 tens as 1 hundred.

3. Add the hundreds: 6. Write 6. Check by estimating.

Add

```
              1              1 1            1 1
   378      3 7 8          3 7 8          3 7 8         400
 + 295    + 2 9 5        + 2 9 5        + 2 9 5       + 300
              3              7 3          6 7 3    is close to   700
```

is close to

Calculator Solution of Example

Press Keys: C 3 7 8 + 2 9 5 = *

Answer: 673.

*Note: A calculator displays an answer only after you press =.

Practice

▮ Add. Check by estimating.

```
      1
1.    38      40        57          41          79          99
    + 19    + 20      + 38        + 26        + 74        + 35
      57      60


      1
2.    272     300       329         490         626         533
    + 167   + 200     + 180       + 264       + 293       + 272
      439     500


      1       1
3.  2,473   2,500     1,831       3,478       5,957       5,500
    + 915   + 900     + 846     + 1,720     + 2,612     + 3,750
    3,388   3,400


      1
4.  $6.75                         $4.36       $7.64       $8.19
    + 3.19                      + 3.29      + 6.75      + 5.90
    $9.94
```

Place a decimal point in the answer to separate dollars and cents.

22

Math Tip

Estimates can differ in value. When you are estimating by rounding, the value of an estimate depends on the place value to which the numbers are rounded:

Exact Answer	Estimate 1	Estimate 2	Estimate 3
468	470	500	470 (nearest 10)
+ 209	+ 210	+ 200	+ 200 (nearest 100)
677	680	700	670
	Rounded to nearest 10	Rounded to nearest 100	Mixed rounding

As you can see, there are many ways to round to find an estimate.

5.

586	845	1,129	$1,853	2,493
+ 375	+ 796	+ 684	+ 759	+ 1,728

6.

182	$347	808	3,500	$4,687
75	89	437	753	2,158
+ 47	+ 76	+ 96	+ 467	+ 1,653

Add. As your first step, line up the digits: ones digit under ones digit and so on.

7. 157 + 89 = $354 + $298 =   1,895 + 762 =

8. 1,308 + 875 = $3,375 + $2,836 =   12,654 + 8,682 =

Use addition to solve each word problem.

9. The distance from Seattle to Portland is 179 miles, while the distance from Portland to Eugene is 114 miles. Driving through Portland, how far is Eugene from Seattle?

10. Estimate the total cost of a $48 set of tools, a $33.88 wheelbarrow, and a $279.95 lawn mower. *Any reasonable estimate is acceptable.*

Subtracting Whole Numbers

To subtract two numbers
- subtract from right to left, starting with the ones column
- regroup as needed

Math Tip

When subtracting, **regrouping** can also be called *borrowing* or *renaming*.

Subtract	1. Regroup 1 ten: 7 → 6.	2. Add 10 to the 4 ones: 10 + 4 = 14.	3. Subtract.	Check by estimating.
74 − 39	6 7̶4 − 39	6 14 7̶4̶ − 39	6 14 7̶4̶ − 39 35	is close to 70 − 40 → 30

Use estimation or a calculator to check your answers.

Calculator Solution of Example
Press Keys: **C 7 4 − 3 9 =**
Answer: **35.**

Practice

▮ Subtract. Check by estimating.

1.
```
   67      70      86        67       154       286
 − 34    − 30    − 55      − 57      − 43      − 75
 ────    ────
   33      40
```

2.
```
  2 14
  3̶4̶2    350     478       549       274       856
 − 160   − 150   − 188     − 264     − 182     − 376
 ─────   ─────
  182     200
```

3.
```
    4 18
  $7.5̶8̶    [Place a decimal point in     $7.90     $8.94     $10.37
 − 3.49    the answer to separate       − 3.28    − 5.77    − 8.09
 ──────    dollars and cents.]
  $4.09
   ↑
```

4.
```
  $5.76          $6.47     $8.50     $19.25     $34.75
 − 3.83         − 2.95    − 5.70    − 12.65    − 25.93
```

24

Example 1. 521 − 179			**Example 2.** 400 − 156	

Example 1. 521 − 179

Step 1.
```
  1 11
  5 2 X
-  1 7 9
```

Step 2.
```
  4 11 11
  5 2 X
-  1 7 9
  3 4 2
```

Example 2. 400 − 156

Step 1.
```
  3 10
  4 0 0
-  1 5 6
```

Step 2.
```
      9
  3 10 10
  4 0 0
-  1 5 6
  2 4 4
```

5.
```
 5 16 14
  6 7 4      700      375        203        634        748
- 2 9 8    - 300    - 176      - 145      - 276      - 389
  3 7 6      400
```

6.
```
  3 0 0      500      412      1,600      1,922
-  1 7 5    - 328    - 153     -  545     -  746
```

7.
```
 $5.00     $6.00     $7.25     $14.00     $35.10
- 2.75    - 3.85    - 2.79     -  6.39    - 11.58
```

■ **Line up the digits and subtract.**

8. 762 − 588 = $12.40 − $8.94 =   1,308 − 948 =

■ **Use subtraction to solve each word problem.**

9. The price of a $314 television is being reduced to $225 during the Memorial Day sale. How much can Gloria save by buying the set at this sale?

11. Find both an estimate and the exact change for Vernon's purchase. Vernon pays with a $50 bill for:

$19.79

Estimate: _____ Exact Change: _____

10. Ervin weighs 79 pounds more than his wife, Rosey. What is Rosey's weight if Ervin tips the scales at 216 pounds?

Problem Solver: Understanding Set-Up Questions

Some test questions, called **set-up questions**, do not ask you to solve a problem in the usual way. Instead, a set-up question asks you to choose a numerical expression or an equation that shows how to compute the correct answer.

Example 1

Shelley has driven 75 miles of the 340-mile distance between Springfield and Oakridge. Which expression at right shows how many miles she has left to drive?

  (1) $340 + 75$
  (2) $75 + 340$
  (3) $340 - 75$
  (4) $75 - 340$
  (5) $340 \div 75$

You think: How much is 340 take away 75?

You choose the expression that represents this subtraction.

The correct answer choice is **(3) 340 − 75.**

Example 2

For the three-day weekend sale, the price of a toaster has been reduced by $5.89. The sale price is $17.60. Which equation at right tells how to calculate the original price (p)?

  (1) $p = \$17.60 - \$5.89$
  (2) $p = \$17.60 + \$5.89$
  (3) $p = \$17.60 + \$11.71$
  (4) $p = \$17.60 - \$11.71$
  (5) $p = \$17.60 \times \$5.89$

You think: The original price (p) is $17.60 plus $5.89.

You choose the equation that represents this addition.

The correct answer choice is **(2) p = $17.60 + $5.89.**

As you see, solving a set-up question involves identifying a numerical expression that represents information given in a sentence or phrase. Here are some more examples.

Sentence or Phrase	Numerical Expression
What is $20 minus $14.95?	$\$20.00 - \$14.95$
How much smaller is 87 than 102?	$102 - 87$
What is the sum of 945 and 893?	$945 + 893$
By how much is $48 more than $31?	$\$48 - \$31$
124 feet decreased by 75 feet	$124 - 75$
the total of $8.46 and $5.76	$\$8.46 + \$5.76$
1,378 pounds increased by 479 pounds	$1{,}378 + 479$

Practice

■ Write a numerical expression for each sentence or phrase.

Sentence or Phrase	Numerical Expression
1. What is $25.00 minus $17.98?	_____
2. Find the sum of 2,305 and 1,987.	_____
3. How long is 62 inches added to 97 inches?	_____
4. $1.98 taken away from $5.00	_____
5. $4.88 more than $2.19	_____
6. 1,345 pounds reduced by 277 pounds	_____

■ Choose the correct expression or equation to solve each problem.

7. Mona made 47 deliveries last month and 53 this month. Which expression shows how many more deliveries Mona made this month than last month?

(1) 47 + 53
(2) 53 − 47
(3) 47 − 53

8. Jonas earned $75 on Monday, $86 on Tuesday, and $105 on Wednesday. Which expression shows the total amount Jonas earned last week?

(1) $105 − $86 − $75
(2) $105 + $86 − $75
(3) $105 + $86 + $75

9. Emma paid $1,850 more for her car than she sold it for. If she sold it for $2,375, which equation represents the amount (a) that Emma paid for her car?

(1) $a = \$2,375 + \$1,850$
(2) $a = \$2,375 - \$1,850$
(3) $a = \$1,850 - \$2,375$

10. In Hank's pocket are several coins: two quarters, three dimes, and four nickels. Which equation tells the total amount (t) of money in cents in Hank's pocket?

(1) $t = 25¢ + 30¢ + 20¢$
(2) $t = 50¢ + 40¢ + 15¢$
(3) $t = 50¢ + 30¢ + 20¢$

11. Carmen paid for a $58.88 wheelbarrow by cashing a paycheck for $203.64. Which expression gives the *best estimate* of the change Carmen should receive?

(1) $200 − $60
(2) $200 + $60
(3) $200 − $50

12. After driving 278 miles, Tuong sees a sign saying "Chicago 412 miles." Which equation gives the *best estimate* of the total miles (m) Tuong will have driven by the time he reaches Chicago?

(1) $m = 400 - 300$
(2) $m = 400 + 200$
(3) $m = 400 + 300$

Many word problems (and problems involving tables, graphs, and maps) contain more numbers than you need to answer a specific question. For these problems, your first step is to decide which numbers are important.

Necessary information includes only those numbers needed to answer a specific question.

Extra information includes those numbers not needed to answer a specific question.

Example 1

Henri planned to spend no more than $35 for a new shirt and a new belt. If the shirt cost $19.95 and the belt cost $12.89, how much did Henri spend in all?

Necessary information: $19.95, $12.89

Extra information: $35

The question asks only how much Henri spent. His spending limit of $35 is not needed to answer this question.

Item Sets

An **item set** is a group of two or more problems that are based on the same passage or graphic. Examples 2–3 are based on the following passage.

> Woody is buying a used pickup that has a sticker price of $5,150. The dealer agreed to lower the price an extra $500 and to give Woody $1,500 for his old car as a trade-in. Woody will pay an additional down payment of $850. For the balance still owing, Woody will take out a loan with the dealer. He will make 12 monthly payments of $206.50 each.

Example 2

What price is Woody paying for the pickup before the amount of the trade-in is deducted?

Necessary information: $5,150 and $500. The price is determined by subtracting $500 from $5,150. Answer: $4,650

Extra information: $1,500, $850, 12, and $206.50

Example 3

How much of a loan will Woody take out with the dealer?

Necessary information: $5,150, $500, $1,500, and $850. The loan amount is determined by subtracting each of the three amounts $500, $1,500, and $850 from the sticker price of $5,150. Answer: $2,300.

Extra information: 12, $206.50

Practice

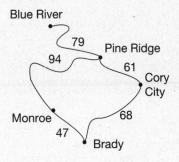

Blue River

79 Pine Ridge

94 61 Cory City

Monroe 68

47 Brady

All distances are in miles.

■ **Problems 1–4 are based on the map at right.**

1 How many miles is Monroe from Cory City if you take the route through Brady?

2. How much farther is Pine Ridge from Monroe than Pine Ridge is from Cory City?

3. How far is Blue River from Brady if you take the route through Monroe?

4. To determine which of the two routes from Brady to Blue River is the shorter, which of the following distances is *not* needed?

   (1) 47
   (2) 61
   (3) 68
   (4) 79
   (5) 94

■ **Problems 5–8 refer to the following passage.**

Leona is trying to decide where to take her daughter Amy for lunch. At Burger Village, children's burgers cost $1.19, french fries cost $0.89, and soft drinks cost $0.79. For herself, Leona would buy the Village Luncheon Special for $3.95. At The Sandwich Shop, a "Kid's Meal" costs $2.19 and includes a hamburger, french fries, and a soft drink. At The Sandwich Shop, Leona's meal costs $4.79.

5. If Amy has a hamburger, fries, and a soft drink, how much will Leona spend for Amy's meal at Burger Village?

6. If Amy has a hamburger, fries, and a soft drink, which expression gives the best estimate of the total cost of lunch if Amy and Leona eat at The Sandwich Shop?

   (1) $1.00 + $4.00
   (2) $2.00 + $4.00
   (3) $2.00 + $5.00
   (4) $3.00 + $5.00
   (5) $3.00 + $6.00

7. How much more would Leona spend on her own meal at The Sandwich Shop than at Burger Village?

8. Which amount is *not* needed to determine how much more Amy's meal costs at Burger Village than at The Sandwich Shop?

   (1) $0.79
   (2) $0.89
   (3) $1.19
   (4) $2.19
   (5) $3.95

Multiplying Whole Numbers

To multiply two numbers
- multiply from right to left, starting with the ones
- regroup as needed
- write zeros as placeholders if you find it helpful when writing partial products

Use estimation or a calculator to check your answers.

Math Tip

Remember:

- zero times a number is zero
- one times a number is that number

$7 \times 0 = 0 \qquad 7 \times 1 = 7$

Multiply

$$\begin{array}{r} 3\,2\,6 \\ \times \quad 4\,8 \end{array}$$

1. Multiply 326 by 8.*
   Write *2608*, writing 8 in the ones place.

$$\begin{array}{r} 2\,4 \\ 3\,2\,6 \\ \times \quad 4\,8 \\ \hline 2\,6\,0\,8 \end{array}$$

1st partial → product

2. Multiply 326 by 4.*
   Write *1304*, writing 4 in the tens place.

$$\begin{array}{r} 1\,2 \\ 3\,2\,6 \\ \times \quad 4\,8 \\ \hline 2\,6\,0\,8 \\ 1\,3\,0\,4\,0 \end{array}$$

2nd partial → product   ← place-holding zero

3. Add the two partial products.

$$\begin{array}{r} 3\,2\,6 \\ \times \quad 4\,8 \\ \hline 2\,6\,0\,8 \\ 1\,3\,0\,4\,0 \\ \hline 1\,5,6\,4\,8 \end{array}$$

*Note: Each partial product has its own carried digits. It is a good idea to write carried digits lightly in pencil. Then you can erase them after you do each multiplication.

Calculator Solution of Example
Press Keys: (C) (3) (2) (6) (×) (4) (8) (=)
Answer: 15648. *

*Note: Most calculators do not display a comma to separate digits into groups of three.

Practice

Multiply. Check by estimating.

1.
$$\begin{array}{r} 23 \\ \times\ 3 \end{array}$$
$$\begin{array}{r} 56 \\ \times\ 5 \end{array}$$
$$\begin{array}{r} 38 \\ \times\ 7 \end{array}$$
$$\begin{array}{r} 51 \\ \times\ 7 \end{array}$$
$$\begin{array}{r} 74 \\ \times\ 9 \end{array}$$

2.
$$\begin{array}{r} 4\,3 \\ \$5.75 \\ \times\ 6 \\ \hline \$34.50 \end{array}$$

Place a decimal point in the answer to separate dollars and cents.

$$\begin{array}{r} \$7.25 \\ \times\ 8 \end{array}$$
$$\begin{array}{r} \$12.50 \\ \times\ 5 \end{array}$$
$$\begin{array}{r} \$25.75 \\ \times\ 9 \end{array}$$

3.
$$\begin{array}{r} 34 \\ \times\ 25 \end{array}$$
$$\begin{array}{r} 46 \\ \times\ 32 \end{array}$$
$$\begin{array}{r} 53 \\ \times\ 47 \end{array}$$
$$\begin{array}{r} 28 \\ \times\ 19 \end{array}$$
$$\begin{array}{r} 79 \\ \times\ 65 \end{array}$$

4.

$7.26	$8.47	638	750	389
× 44	× 23	× 53	× 61	× 75

▪ **In Problems 5–6, a shortcut is shown that you can use when the bottom number contains a zero. Use a place-holding 0 (as shown in the shortcut) instead of writing a partial product that contains only 0s (000).**

Shortcut   Long Way

5.

$6.43	$6.43	$8.57	739	382
× 20	× 20	× 40	× 60	× 70
$128.60	000			
	128 6			
	$128.60			

Shortcut   Long Way

6.

864	864	675	396	708
× 605	× 605	× 507	× 306	× 403
4320	4320			
518400	0000			
522,720	518 400			
	522,720			

▪ **Line up the numbers and multiply.**

7. 79 × 61 =   85 × 43 = $5.67 × 25 = $14.62 × 50 =

▪ **Use multiplication to solve each word problem.**

8. What is the cost of 19 gallons of gas priced at $1.27 per gallon?

9. Kate's car gets 36 miles per gallon during highway driving. On a cross-country trip, how far can Kate's car go on a full tank of 18 gallons?

10. Find both an estimate and the exact price of the purchase shown below.

Wally buys 28 gallons at $10.39 per gallon.

Estimate:_____
Exact Price:_____

Dividing Whole Numbers

Dividing by a One-Digit Number

Division is a 5-step process that combines many of the skills you now have. For longer problems, you may need to repeat these steps.
- Think of a long division problem as being made up of several short divisions.
- When division does not end in zero, write *r* and the amount left over to show the remainder.

5 Division Steps

- Divide; then write a digit in the answer.
- Multiply.
- Subtract.
- Compare.
- Bring down the next digit.

Divide: $4\overline{)95}$

1st division: Divide 4 into 9	**2nd division: Divide 4 into 15**
Step 1. Divide: $9 \div 4 = 2$ Write *2* above the 9.	*Step 1.* Divide: $15 \div 4 = 3$ Write *3* above the 5.
Step 2. Multiply: $2 \times 4 = 8$	*Step 2.* Multiply: $3 \times 4 = 12$
Step 3. Subtract: $9 - 8 = 1$	*Step 3.* Subtract: $15 - 12 = 3$
Step 4. Compare: $1 < 4$	*Step 4.* Compare: $3 < 4$
Step 5. Bring down the 5.	*Step 5.* There is no other digit to bring down. Write *3* as the remainder.

$$\begin{array}{r} 2 \\ 4\overline{)95} \\ -8\downarrow \\ \hline 15 \end{array}$$

$$\begin{array}{r} 23\ r3 \\ 4\overline{)95} \\ -8 \\ \hline 15 \\ -12 \\ \hline 3 \end{array}$$

Calculator Solution of Example
Press Keys: C 9 5 ÷ 4 =
Answer: 23.75 *

*Note: A calculator displays a remainder as a decimal fraction. You'll study decimals in Unit 5.

Practice

Divide. Check by estimating.

1. $\begin{array}{r} 9\ r3 \\ 5\overline{)48} \\ -45 \\ \hline 3 \end{array}$ $\begin{array}{r} 10 \\ 5\overline{)50} \end{array}$ $6\overline{)25}$ $7\overline{)54}$ $3\overline{)19}$ $8\overline{)60}$

2. $\begin{array}{r} 54 \\ 6\overline{)324} \\ -30 \\ \hline 24 \\ -24 \\ \hline 0 \end{array}$ $\begin{array}{r} 50 \\ 6\overline{)300} \end{array}$ $4\overline{)336}$ $5\overline{)295}$ $4\overline{)892}$ $3\overline{)728}$

3. $3\overline{)\$8.25}$ ↓ | Place a decimal point in the answer. $2\overline{)\$7.48}$ $5\overline{)\$9.25}$ $6\overline{)\$25.50}$

Using Zero as a Placeholder

Use 0 as a placeholder in the quotient each time you
- divide into 0
- divide into a number that is smaller than the divisor

Example 1. $3\overline{)609}$
1st division

$$\begin{array}{r} 2 \\ 3\overline{)609} \\ -6 \\ \hline 00 \end{array}$$

2nd division

$$\begin{array}{r} 203 \\ 3\overline{)609} \\ -6\downarrow \\ \hline 009 \\ -9 \\ \hline 0 \end{array}$$

Divide 3 into 00: 00 ÷ 3 = 0
Write *0* in the quotient.
Bring down 9; then divide 3 into 9.
Final 0 shows no remainder.

Example 2. $4\overline{)817}$
1st division

$$\begin{array}{r} 2 \\ 4\overline{)817} \\ -8\downarrow \\ \hline 01 \end{array}$$

2nd division

$$\begin{array}{r} 204\ r1 \\ 4\overline{)817} \\ -8\downarrow \\ \hline 017 \\ -16 \\ \hline 1 \end{array}$$

4 won't divide into 1, so
write *0* in the quotient.
Bring down 7; then divide 4
into 17. Remainder = 1.

4. $\begin{array}{r} 70 \\ 3\overline{)210} \\ -21 \\ \hline 000 \end{array}$ $5\overline{)250}$ $7\overline{)\$490}$ $6\overline{)4,200}$ $5\overline{)\$35,000}$

5. $\begin{array}{r} \$2.02 \\ 4\overline{)\$8.08} \\ -8 \\ \hline 008 \\ -8 \\ \hline 0 \end{array}$ $3\overline{)906}$ $2\overline{)\$6.08}$ $4\overline{)1,208}$ $6\overline{)1,206}$

6. $3\overline{)614}$ $5\overline{)549}$ $6\overline{)\$6.24}$ $8\overline{)\$1,648}$ $9\overline{)2,719}$

Dividing by a Two-Digit Number

You use the same 5 steps to divide by a two-digit number that you use to divide by a one-digit number. Estimating can help with each division.

Divide: $17\overline{)795}$

1st division: Divide 17 into 79	2nd division: Divide 17 into 115,
Step 1. Divide: $79 \div 17 = ?$ Estimate: $80 \div 20 = 4$ Try 4.	*Step 1.* Divide: $115 \div 17 = ?$ Estimate: $120 \div 20 = 6$ Try 6.
Step 2. Multiply: $17 \times 4 = 68$ Write 4 above 9.	*Step 2.* Multiply: $17 \times 6 = 102$ Write 6 above 5.
Step 3. Subtract: $79 - 68 = 11$	*Step 3.* Subtract: $115 - 102 = 13$
Step 4. Compare: $11 < 17$ (4 is correct because $68 < 79$, and $11 < 17$.*)	*Step 4.* Compare: $13 < 17$ (6 is correct because $102 < 115$ and $13 < 17$.*)
Step 5. Bring down the 5.	*Step 5.* There is no other digit to bring down. Remainder = 13.

1st division:
$$\begin{array}{r} 4 \\ 17\overline{)795} \\ -68 \\ \hline 115 \end{array}$$

2nd division:
$$\begin{array}{r} 46\ r13 \\ 17\overline{)795} \\ -68 \\ \hline 115 \\ -102 \\ \hline 13 \end{array}$$

*If these two comparison conditions are not met, choose a new quotient digit and try again.

Practice

▨ **In rows 7–8, check to see if each quotient digit is correct. Cross out each incorrect digit; then write the correct digit above it.**

7.
$$\begin{array}{r} 2 \\ 18\overline{)56} \\ - \\ \hline \end{array} \qquad \begin{array}{r} 3 \\ 32\overline{)97} \\ - \\ \hline \end{array} \qquad \begin{array}{r} 5 \\ 24\overline{)119} \\ - \\ \hline \end{array} \qquad \begin{array}{r} 5 \\ 37\overline{)219} \\ - \\ \hline \end{array} \qquad \begin{array}{r} 6 \\ 28\overline{)210} \\ - \\ \hline \end{array}$$

8.
$$\begin{array}{r} \text{Check} \\ 18 \\ 28\overline{)549} \\ -28 \\ \hline 269 \\ - \\ \hline \end{array} \qquad \begin{array}{r} \text{Check} \\ 23 \\ 19\overline{)422} \\ -38 \\ \hline 42 \\ - \\ \hline \end{array} \qquad \begin{array}{r} \text{Check} \\ 24 \\ 33\overline{)823} \\ -66 \\ \hline 163 \\ - \\ \hline \end{array} \qquad \begin{array}{r} \text{Check} \\ 45 \\ 16\overline{)750} \\ -64 \\ \hline 110 \\ - \\ \hline \end{array} \qquad \begin{array}{r} \text{Check} \\ 32 \\ 22\overline{)718} \\ -66 \\ \hline 58 \\ - \\ \hline \end{array}$$

▨ **Divide. Check on scratch paper by estimating.**

9. $27\overline{)243} \qquad 34\overline{)204} \qquad 18\overline{)144} \qquad 21\overline{)153} \qquad 38\overline{)347}$

34

10. $19\overline{)647}$ $\qquad$ $31\overline{)589}$ $\qquad$ $22\overline{)879}$ $\qquad$ $43\overline{)907}$ $\qquad$ $54\overline{)849}$

11. $15\overline{)1,650}$ $\qquad$ $26\overline{)4,654}$ $\qquad$ $35\overline{)8,295}$ $\qquad$ $19\overline{)8,320}$ $\qquad$ $46\overline{)12,675}$

■ **Rewrite each problem with a division bracket; then divide.**

12. $518 \div 7 =$ $\qquad$ $605 \div 5 =$ $\qquad$ $941 \div 8 =$ $\qquad$ $1,654 \div 9 =$

13. $234 \div 18 =$ $\qquad$ $452 \div 24 =$ $\qquad$ $2,853 \div 13 =$ $\qquad$ $6,050 \div 29 =$

■ **Use division to solve the following problems.**

14. When a deck of 52 cards is dealt to 4 people, how many cards does each person receive?

15. If Yolanda can pack 36 calculators into each shipping box, how many boxes will she need to send 2,448 calculators?

16. On the average, it takes Lydia 8 minutes to check a charge-card application form. How many complete forms can Lydia check during her final hour of work?

17. Kirby Bus Company is busing 329 students to a music conference. If each bus holds a maximum of 38 students, how many buses will be needed?

Problem Solver: Solving Multi-Step Problems

In a multi-step problem, you use two or more operations to compute an answer. The key to solving a multi-step problem is to think of it as two or more one-step problems.

Example 1

Maria and three friends agreed to evenly split the cost of dinner. The bill included $21.75 for pizzas, $7.50 for salads, and $8.75 for soft drinks. How much was Maria's share of the bill?

You think: First, I need to know the total cost of the meal.
Then, I divide this cost by 4 to find Maria's share.

Step 1. Letting c stand for *cost of meal*, you can write:
$c = \$21.75 + \$7.50 + \$8.75 = \$38.00$
Step 2. Letting M stand for *Maria's share*, you write:
$M = \$38.00 \div 4 = \mathbf{\$9.50}$

Answer: Maria's share is $9.50.

> **Math Tip**
>
> Although it's not necessary, writing an equation for each step is a good idea. Many students, though, prefer to write just the correct numerical expression.

In many multi-step problems, there is more than one correct way to compute the answer.

Example 2

Phil earns $7.50 for each hour he works on Thursday and Friday. How much will Phil earn on these two days when he works for 8 hours on Thursday and 6 hours on Friday?

You may think: First, I need to find how many total hours he'll work.
Then, I multiply the total hours by $7.50.
If h stands for *total hours worked*, and a for *total amount earned*, you can write the two steps as:

Step 1. $h = 8 + 6 = 14$ hours
Step 2. $a = \$7.50 \times 14 = \mathbf{\$105.00}$

Or you may think: First, I need to find out how much he earned each day.
Then, I add the two amounts to find the total.

If T stands for *amount earned on Thursday*, F for *amount earned on Friday*, and a for *total amount earned*, you write:

Step 1. $T = \$7.50 \times 8$ and $F = \$7.50 \times 6$
      $= \$60.00$ $= \$45.00$
Step 2. $a = \$60.00 + \$45.00 = \mathbf{\$105.00}$

Answer: Phil will earn $105. Either solution gives the correct answer.

Practice

■ **Write a numerical expression for each sentence or phrase.**

Sentence or Phrase   **Numerical Expression**

1. What is the product of 123 times 17?   _____

2. How many times does 9 go into 236?   _____

3. What distance is twice 748 miles?   _____

4. $134.88 split 6 ways   _____

5. the quotient of 3,540 divided by 35   _____

■ **Solve each problem below.**

6. Kira baked cookies for her daughter's second-grade class. From the 93 cookies she baked, she kept 24 home for her family. She took the rest to school. If there are 23 children in the class, how many cookies will each child get?

7. For his car tune-up business, Herman bought 12 cases of motor oil. Each case contains 24 quarts of oil. If each tune-up takes an average of 4 quarts, how many tune-ups can Herman do before he needs more oil?

8. Roberto can carry 27 bales of hay in his pickup. He can carry an additional 15 bales in a small trailer attached to the pickup. How many bales can Roberto move in 7 trips if he uses both pickup and trailer?

9. Thomas planned to spend $35.00 at The Man's Store. After buying a shirt for $14.95 and a tie for $9.99, how much did Thomas have left to spend?

10. When Jerry bought his Honda, the mileage indicator read 18,146 miles. Now, after owning the car for 1 year, Jerry notices that the indicator reads 33,062 miles. On the average, how many miles has Jerry driven each month?

11. Manny and three friends agreed to split the cost of dinner. They had a pizza for $14.70, soft drinks for $5.50, and salads for $8.40. What was Manny's share of the bill?

12. As part of a promotion, Lita gave away 60 balloons every day during the 31 days in July. At the end of the month, she still had 132 balloons left over. Knowing this, figure out how many total balloons Lita started out with.

13. Each Monday through Friday, Marlo delivers a newspaper to each of her 74 customers. Last weekend she delivered a total of 114 additional papers. How many papers did Marlo deliver last week?

14. Look at the map below. How many more miles do you fly if you connect at Chicago on a trip from Denver to New York than if you fly direct?

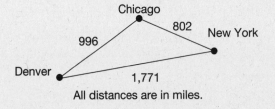

All distances are in miles.

Problem Solver: Choosing the Correct Expression

In the previous problem solver, you saw how a multi-step problem is solved by breaking it down into simpler one-step problems. Each step has its own numerical expression.

On this page, you'll learn how to write a **single numerical expression** to represent all steps needed to solve a multi-step problem. The key is to learn the correct order in which to solve these expressions.

Order of Operations

Parentheses ( ) are used to indicate a single quantity that is to be multiplied or divided. When you use parentheses, you do not need to write a multiplication sign.

To find the value of a numerical expression, follow these steps:
• Do any operation indicated within parentheses first.
• Next, do any operation on the top or bottom of a division bar before dividing.
• Then, starting at the left, do all multiplication and division.
• Last, starting at the left, do all addition and subtraction.

Numerical Expression		Finding the Value
$3 \times 4 + 9$	*Step 1.* Multiply: $3 \times 4 = 12$	*Step 2.* Add: $12 + 9 = 21$
$14 - 24 \div 8$	*Step 1.* Divide: $24 \div 8 = 3$	*Step 2.* Subtract: $14 - 3 = 11$
$\$20 - (\$5 + \$3)$	*Step 1.* Add: $\$5 + \$3 = \$8$	*Step 2.* Subtract: $\$20 - \$8 = \$12$
$\dfrac{25 - 9}{8}$ ← division bar	*Step 1.* Subtract: $25 - 9 = 16$	*Step 2.* Divide: $16 \div 8 = 2$
$4(12 - 7)$	*Step 1.* Subtract: $12 - 7 = 5$	*Step 2.* Multiply: $4 \times 5 = 20$
$(\$19 + \$9) \div 7$	*Step 1.* Add: $\$19 + \$9 = \$28$	*Step 2.* Divide: $\$28 \div 7 = \$4$

Write two numerical expressions that will solve the problem below.
Jamie earns $6.75 for each hour she works on the weekend. If she worked for 8 hours on Saturday and 6 hours on Sunday, how much did she earn over the weekend?

The two correct numerical expressions for the solution of this problem are
1. total earned = pay per hour *times* total hours worked
   = $6.75(8 + 6)
   or
2. total earned = Saturday's earnings *plus* Sunday's earnings
   = ($6.75 × 8) + ($6.75 × 6)

Each numerical expression gives an answer of $94.50.

Practice

■ **Write a numerical expression for each sentence or phrase below.**

Sentence or Phrase	Numerical Expression
1. How much is $18.99 added to the product of $5.50 times 7?	_____
2. Find the sum of 345 and 125; then divide the total by 3.	_____
3. What is $45 reduced by the quotient of $60 divided by 2?	_____
4. the sum of 37 plus 29 plus 14; the total multiplied by 4	_____
5. $100 divided by the sum of 2 + 4	_____

■ **Find the value of each expression below.**

6. $12 \times 4 - 16$ $9 \times 8 + 24$ $36 - 12 \div 2$

7. $\dfrac{35 - 19}{4}$ $5(15 - 9)$ $(\$24 + \$13) - (\$12 + \$9)$

■ **For problems 8–9, choose the one correct expression.**

8. Amy gave a clerk $20 for a $10.98 gallon of paint and one paintbrush. The paint cost $10.98, the brush cost $4.89, and the sales tax was $0.88. Which expression shows how much change Amy should get?

   (1) ($10.98 + $4.89 + $0.88) − $20
   (2) $20 − $10.98 − $4.89 + $0.88
   (3) $20 − ($10.98 + $4.89 + $0.88)
   (4) $20 − ($10.98 + $4.89 − $0.88)
   (5) $20 + $10.98 + $4.89 − $0.88

9. Sandi and two friends agreed to split the cost of lunch. Hamburgers cost $6.57, soft drinks $4.50, and salads $5.25. Which expression shows Sandi's share?

   (1) ($6.57 + $4.50 + $5.25) 2
   (2) 3($6.57 + $4.50 + $5.25)
   (3) $\dfrac{\$6.57}{3 + (\$4.50 + \$5.25)}$
   (4) $\dfrac{\$6.57 + \$4.50 + \$5.25}{3}$
   (5) 2($6.57 + $4.50 + $5.25)

■ **For problem 10, there are two correct expressions for the answer. Choose these two expressions.**

10. Each day he works overtime, Daniel earns $18 more than his regular daily pay of $64. Last week, he worked overtime 2 out of 5 days. How much did Daniel earn last week?

   (1) 3($64 × $18) + 2($64)
   (2) 5($64 + $18)
   (3) $64 × 5 + $18 × 2
   (4) $64 × 3 + $18 × 2
   (5) 2($64 + $18) + 3($64)

Data Highlight: Finding Typical Values

Look at the list of pizza prices shown at right. Suppose a friend, Dana, asks you, "What is the typical price of a large pizza in Albany?"

What would you tell Dana? $13, $14, $15, $16, or some other amount? Just what is *typical?*

In math, there are three main ways to give a **typical value**: as a **mean**, a **median**, or a **mode**.

Pizza Prices in Albany	
Joey's Large Pizza	$14
Mia's Large Pizza	$13
Tino's Large Pizza	$16
Little Italy's Large Pizza	$14
Franko's Large Pizza	$15

Mean

Mean is another word for *average.* To find the average of a set (group) of numbers
- add the numbers in the set
- divide this sum by the number of numbers in the set

Usually, when people talk about average, they're referring to mean.

Note: In most cases, the mean is not equal to any number in a set. However, the mean is often close in value to the middle number of the set.

To find the mean price of a large pizza in Albany, follow these steps:

Step 1. Add

$14
13
16
14
+ 15
$72

Step 2. Divide

$14.40 ← mean
5) $72.00 ← sum
        ↑
        number
        in set

Mean = $14.40

Median

The *median* is the middle value of a set of numbers, arranged from least to greatest value.
- If a set contains an odd number of numbers, the median is the middle number.
- If a set contains an even number of numbers, the median is the average of the two middle numbers.

To find the median price of a large pizza in Albany, do the following.

Step 1. Arrange the prices in order, from least to greatest.
$13, $14, $14, $15, $16
            ↑ middle value

Step 2. Since there is an odd number of prices, the median is the middle value.

Median = $14

Mode

The *mode* of a set of numbers is the number that appears the most times in the set. If no number appears more than once, a set has no mode.

The mode of the pizza prices is $14, since $14 appears more than any other price.

Mode = $14

Practice

■ Find the *mean, median,* and *mode* for each set of numbers in problems 1–2.

1. Shoe Prices

$15
$18
$20
$26
$26

Mean: _____

Median: _____

Mode: _____

2. House Sizes (square feet)

1,400
1,400
1,550
1,850
2,000

Mean: _____

Median: _____

Mode: _____

3. The 4 new models at Al's Car City have the following mileage ratings.

	City	Highway
Model DX	33	38
Model DXi	28	35
Model LX	26	31
Model SX	17	24

a) What is the average city mileage rating of the four models listed?

b) What is the average highway mileage rating of the four models listed?

c) What is the median city mileage rating of the four models listed? (Remember: For an even number, the median is the average of the two middle values.)

4. Below is information on shoes being sold at Family Shoe Store.

	Running Shoes		
Model	Selling Price	Store's Cost	Profit
	(a)	(b)	(a - b)
A	$60.00	$40.00	_____
B	$52.00	$36.00	_____
C	$48.00	$34.00	_____
D	$40.00	$25.00	_____

a) Compute the profit made on each model.

b) Compute the mean profit of the four models listed.

c) Determine the median profit of the four models listed.

5. Below is a list of Ben's overtime hours.

	April	May	June
Week 1:	12	9	14
Week 2:	10	8	12
Week 3:	8	12	9
Week 4:	10	10	13

Which equation below shows how to compute the average weekly overtime hours (*h*) Ben worked during May?

(1) $h = (12 + 9 + 14) \div 3$

(2) $h = (9 + 8 + 12 + 10) \div 3$

(3) $h = (9 + 8 + 12 + 10) \div 4$

(4) $h = (12 + 9 + 14) \div 4$

(5) $h = 4(9 + 8 + 12 + 10)$

Test Readiness Checkup

Circle each correct answer. Check your answers on page 240; then correct any errors.

1. For July, Francine's employer withheld $278 for federal taxes, $59 for Social Security, and $37 for medical insurance. What is the total of these three deductions?

   (1) $334
   (2) $354
   (3) $374
   (4) $394
   (5) $414

2. The first football game at Franklin High School had an attendance of 5,219. The second game drew only 4,847 fans. How much did the attendance decrease between these two games?

   (1) 282
   (2) 312
   (3) 342
   (4) 372
   (5) 402

3. Erin's Restaurant sold 257 large Cokes over the weekend. If each large cup holds 16 ounces of pop, how many total ounces of Coke did Erin's sell during this two-day period?

   (1) 3,112
   (2) 3,442
   (3) 3,742
   (4) 4,112
   (5) 4,332

4. For a loan of $5,000 from Friendly Finance, Bert would have to pay back a total of $5,679.60 divided into 24 equal monthly payments. What would be Bert's monthly loan payment?

   (1) $236.65
   (2) $243.80
   (3) $250.25
   (4) $263.75
   (5) $271.30

5. The sticker on Phan's new car rates its mileage at 36 miles per gallon for highway driving and 27 miles per gallon in the city. For city driving, how many miles can Phan drive if the tank holds 18 gallons?

   (1) 348
   (2) 396
   (3) 486
   (4) 648
   (5) 972

6. Wanda paid for a $28.99 blouse with a check for $50.00. With the change, she's decided to buy as many $2 bottles of hair shampoo as she can. How many bottles is she able to buy?

   (1) 9
   (2) 10
   (3) 11
   (4) 14
   (5) 25

7. Virginia wants to know how much more a set of 6 walnut chairs costs than a set made up of 4 pecan chairs and 2 maple chairs. Referring to the drawing below, which amount is the best estimate of this difference?

   (1) $60
   (2) $80
   (3) $100
   (4) $120
   (5) $140

| Maple | Oak | Walnut | Pecan |
| $139.49 | $131.29 | $150.89 | $121.79 |

8. A case (12 quarts) of motor oil normally sells for $11.99 but is on sale for $0.79 per quart. The oil company is offering a $1.50-per-case rebate. Suppose Al buys a case at the sale price and sends for the rebate. Which expression shows how much the case will end up costing Al?

(1) $11.99 − $1.50
(2) 12($1.50 − $0.79)
(3) $11.99 − $0.79 × 12
(4) $11.99 − ($0.79 × 12 − $1.50)
(5) $0.79 × 12 − $1.50

9. Over the past 30 months, Bea's weight dropped from 230 to 148. Her goal was to lose 100 pounds and reach an ideal weight of 130. She figures this goal is still 6 months off. Up to now, though, which expression shows Bea's average monthly weight loss?

(1) $\frac{100}{(30 + 6)}$

(2) $\frac{(230 - 148)}{30}$

(3) $\frac{(230 - 130)}{30}$

(4) $\frac{(230 - 130)}{(30 + 6)}$

(5) $\frac{(230 - 148)}{(30 + 6)}$

■ **Problems 10–11 are based on the following passage.**

Monty applied for a job as an administrative assistant with an accounting firm. As part of the application process, he had to take a 16-minute typing test. During the 16 minutes, he typed 1,072 words. Monty got the job at a starting salary of $1,388 per month. After 8 months, Monty got a raise to his present salary of $1,492 per month.

10. During the typing test, how many words per minute did Monty type on the average?

(1) 63
(2) 65
(3) 67
(4) 69
(5) 71

11. Which expression is the best estimate of Monty's earnings during his first six months on the job?

(1) $1,400 × 6
(2) $1,500 × 6
(3) $1,400 × 8
(4) $1,500 × 8
(5) $1,500 × 10

■ **Problems 12–13 refer to the table shown at right.**

12. What is the mean price of homes listed for the city of Lewisburg?

(1) $74,000
(2) $80,900
(3) $83,080
(4) $86,540
(5) $98,000

13. Which equation shows how to compute the median price of homes listed for Marysville?

(1) $p = (\$82,500 + \$87,000) \div 6$
(2) $p = (\$72,000 + \$97,650) \div 6$
(3) $p = (\$82,500 + \$87,000) \div 6$
(4) $p = (\$72,000 + \$97,650) \div 2$
(5) $p = (\$82,500 + \$87,000) \div 2$

HOMES FOR SALE		
Lewisburg	Garrett	Marysville
$68,500	$74,800	$72,000
$74,000	$78,000	$76,500
$80,900	$86,900	$82,500
$94,000	$92,000	$87,000
$98,000	$99,900	$91,800
	$104,000	$97,650

UNIT 3 Numbers Smaller than 1

Parts of a Whole

A number smaller than 1 represents part of a whole. People talk about parts of a whole all the time.
"Drive *three-tenths* of a mile and then turn right."

"You'll need a *quarter* pound of butter for the cake."
"*Forty percent* of the class was absent on Tuesday."

You can divide any whole object—such as a pound or a mile
—or any group, such as a class, into smaller parts.

If you represent the whole object with the number 1, you can
represent any part with a number smaller than 1.

In this chapter, you'll learn the meanings and uses of the three
ways to write numbers smaller than 1.

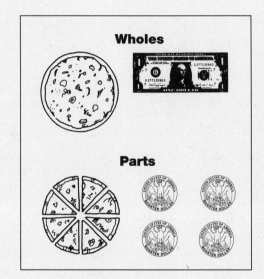

Wholes

Parts

• decimal fractions: 0.3 *or* three-tenths

• proper fractions: $\frac{1}{4}$ *or* one-fourth *or* one-quarter

• percents: 40% *or* 40 percent *or* forty percent

Practice

▨ **Listed below are the most commonly used words standing for parts. Be sure to know the meaning and spelling of each of these words.**

1. Write the correct words in the Skill Check column to name the part shown.

Whole Object	Shaded Amount	Skill Check
a)	one-*half*	1 of 2 parts is called _____
b)	one-*third*	1 of 3 parts is called _____
c)	one-*fourth*	1 of 4 parts is called _____
d)	one-*fifth*	1 of 5 parts is called _____
e)	one-*sixth*	1 of 6 parts is called _____
f)	one-*seventh*	1 of 7 parts is called _____
g)	one-*eighth*	1 of 8 parts is called _____

Whole Object	Shaded Amount	Skill Check
h) ▢▢▢▢▢▢▢▢▢	one-*ninth*	1 of 9 parts is called _____
i) ▢▢▢▢▢▢▢▢▢▢	one-*tenth*	1 of 10 parts is called _____
j) [100 grid]	one-*hundredth*	1 of 100 parts is called _____

100

k) [1,000 grid]	one-*thousandth*	1 of 1,000 parts is called _____

1,000

2. For each statement below, circle the whole that is being discussed and underline the part.

a) "Janice spent almost half an hour on the telephone!"

b) "The bank will loan us ninety percent of the $14,500 we need for the car."

c) "The recipe calls for three-fourths of a pound of beef."

d) "Our train station is only four-tenths of a mile away."

e) "The blouse is on sale for thirty-five percent off the regular price."

f) "The machinist shaved eight-hundredths of an inch off the fitting."

3. Write *d*, *f*, or *p* to indicate whether each of the following is usually expressed as a decimal (d), fraction (f), or percent (p).

_____ a) a number of cents less than one dollar

_____ b) the length of a bolt that's shorter than 1 inch

_____ c) the amount of discount to be given at a store sale

_____ d) the rate of interest that a bank offers on a savings account

_____ e) human body temperature as it's read on a thermometer

_____ f) a small amount of flour that is measured in a measuring cup

Reading and Writing Decimal Fractions

The First Three Decimal Places

The first three decimal places are **tenths, hundredths,** and **thousandths.** Most decimal fractions you'll ever use will have three or fewer digits.

$$. \quad X \qquad X \qquad \quad X$$

tenths ⬏   ↑   ⬑ thousandths
hundredths

Reading Decimal Fractions

- Read a decimal fraction as a number *plus* the place value of the right-hand digit.
- Ignore the leading 0 when reading a decimal fraction.

Example	Number + Place Value	
0.1	**1**	**tenth**
0.5	5	tenths
0.01	**1**	**hundredth**
0.05	5	hundredths
0.23	23	hundredths
0.001	**1**	**thousandth**
0.005	5	thousandths
0.023	23	thousandths
0.150	150	thousandths

Math Tip

Decimal Fraction	Meaning
0.1	1 of 10 equal parts
0.01	1 of 100 equal parts
0.001	1 of 1,000 equal parts

It is common to write a **leading zero** if the whole number is 0. For test practice, we'll write leading zeros from now on.

Math Tip

When writing dollars and cents, write cents in the first two decimal places.

dollars ⬎

$7.29

dimes ⬏ ⬑ pennies
(.1 dollar)   (.01 dollar)

Read the decimal point as the word *and*.

$7.29 is seven dollars *and* 29 cents.

Practice

How is each decimal fraction read in words?

1. 0.3 _____

2. 0.5 _____

3. 0.09 _____

4. 0.20 _____

5. 0.007 _____

6. 0.080 _____

7. 0.075 _____

8. 0.425 _____

For money amounts, write the decimal point as the word *and.*

9. $2.05 _____

10. $12.07 _____

11. $3.88 _____

12. $40.50 _____

Writing Decimal Fractions

To write a decimal fraction
- first, identify the place value of the right-hand digit
- next, write the number so that the right-hand digit is in its proper place; then write 0s as placeholders if necessary

Write 27 hundredths as a decimal fraction. Write *27* so that the 7 ends up in the hundredths place—the second place to the right of the decimal. For practice, write a leading 0. ↓ 0.27 ↑ Write 7 in the hundredths place.	Write 48 thousandths as a decimal fraction. Write *48* so that the 8 ends up in the thousandths place—the third place to the right of the decimal. Write 0 as a placeholder. ↓ 0.048 ↑ Write 8 in the thousandths place.

Practice

Solve.

13. Write each quoted amount on the purchase order at right.

   "Type A, fifty-six thousandths of an inch."

   "Type B, three-tenths of an inch."

   "Type C, twenty-four thousandths of an inch."

   "Type D, two hundred nine thousandths of an inch."

   "Type E, one hundred twenty thousandths of an inch."

   "Type F, seven-hundredths of an inch."

   "Type G, eight-tenths of an inch."

PURCHASE ORDER	
Item Type	**Wall Thickness**
Type A	_____ inch
Type B	_____ inch
Type C	_____ inch
Type D	_____ inch
Type E	_____ inch
Type F	_____ inch
Type G	_____ inch

14. Reporting on a 100-meter race, the announcer told the crowd by how much each runner "just missed the track record." Write these times as decimal fractions on the form at right.

   "Lopez, seventy-eight thousandths of a second."

   "Jenkins, two-tenths of a second."

   "Morris, fifty-five hundredths of a second."

   "Myers, one hundred twelve thousandths of a second."

   "Calder, six-tenths of a second."

   "Nomura, eighty-six hundredths of a second."

RACE RESULTS	
Runner	**Time Off Record**
Lopez	_____ sec.
Jenkins	_____ sec.
Morris	_____ sec.
Myers	_____ sec.
Calder	_____ sec.
Nomura	_____ sec.

Comparing and Ordering Decimals

When you are comparing decimal fractions and **mixed decimals**—a whole number plus a decimal fraction—use the familiar comparison symbols at right (first discussed on page 4).

To make comparison easier, add 0s to give decimal fractions the same number of decimal places.

Adding a zero to the right-hand end of a decimal fraction does not change its value.

Math Tip

$>$ means *is greater than*
$<$ means *is less than*
$=$ means *is equal to*
$\neq$ means *is not equal to*

.5 = .50

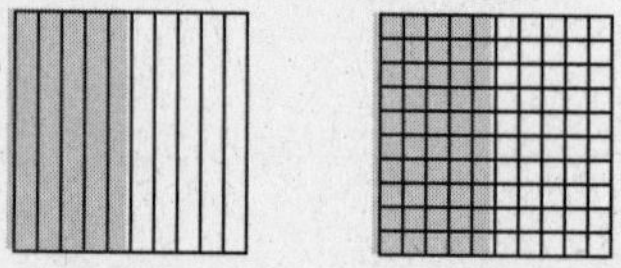

5 tenths = 50 hundredths

Place 0 at the end

0.5 = 0.50

↑ like this

not at the front.

0.5 ≠ ~~0.05~~

↑ not like this

Compare 0.35 and 0.42.

0.[3]2 Because $4 > 3$,
0.[4]2 $0.42 > 0.32$

Compare \$2.65 and \$2.70.

\$2.[6]5 Because $6 < 7$,
\$2.[7]0 \$2.65 < \$2.70
↑ same first digit

Compare 0.18 and 0.187.

Add a 0.
0.1 8[0] Because $0 < 7$,
0.1 8[7] $0.18 < 0.187$
↑↑ same first two digits

Practice

Part A. Write >, <, or = to compare each pair of decimals below.

1. 0.6 ____ 0.8   0.93 ____ 0.9   0.13 ____ 0.24
2. 0.29 ____ 0.290   0.172 ____ 0.208   0.76 ____ 0.547
3. \$0.09 ____ \$0.15   \$0.18 ____ \$0.70   \$0.40 ____ \$.40
4. 3.56 ____ 3.359   1.50 ____ 1.5   12.456 ____ 12.63
5. \$2.45 ____ \$1.98   \$0.89 ____ \$1.09   \$12.45 ____ \$9.99

Part B. For each swimming race recorded below, list the swimmers in order of finish. Remember: The shorter the time, the better the finish.

Race #1 Times	Race #1 Results	Race #2 Times	Race #2 Results
Bill: 48.8 sec.	1st: ______	Jackie: 24.75 sec.	1st: ______
James: 49.08 sec.	2nd: ______	Marina: 23.8 sec.	2nd: ______
Pablo: 48.79 sec.	3rd: ______	Fran: 23.785 sec.	3rd: ______
Jamal: 49.125 sec.	4th: ______	Anna: 24.29 sec.	4th: ______

Life Skill: Reading a Metric Ruler

Probably the most commonly used measuring tool found around home is the **ruler**. Pictured below is a **15-centimeter ruler**. Each centimeter (cm) is divided into 10 **millimeters** (mm): 1 cm = 10 mm.

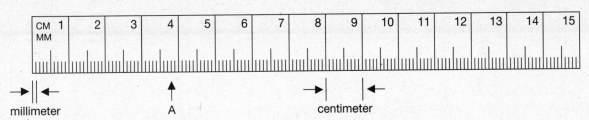

On a centimeter ruler, you read a distance as a number of centimeters plus a number of millimeters. Seeing that 1 mm = 0.1 cm, you can write a distance in two ways:

- As centimeters *and* millimeters: Point A is read as **3 cm 8 mm.**
- As centimeters *only:* Point A is read as **3.8 cm.**

To write a distance in centimeters only, write the number of millimeters as the first number to the right of the decimal point.

Write 5 cm 7 mm as centimeters only.	Write 8.2 cm as centimeters and millimeters.	Write 49 mm in two other ways.
Answer: 5.7 cm	**Answer: 8 cm 2 mm**	**Answer: 4 cm 9 mm and 4.9 cm**

Practice

■ **Solve.**

1. Write 9 cm 6 mm as centimeters only.

   _____

2. Write 3.4 cm as centimeters and millimeters.

   _____

3. Write 138 mm in two other ways.

   cm and mm: _____
   cm only: _____

4. What is the length of each object pictured below? Write your answers on the lines provided.

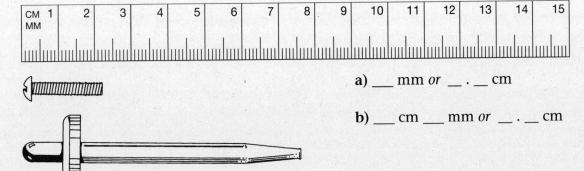

   a) ___ mm *or* ___ . ___ cm

   b) ___ cm ___ mm *or* ___ . ___ cm

Writing Proper Fractions

A proper fraction also stands for part of a whole. A proper fraction is written as one number above a second number.
- The top number, the **numerator**, tells the number of parts you're describing.
- The bottom number, the **denominator**, tells the number of equal parts the whole is divided into.

The whole may be a single object. *or* **The whole may be a group of objects.**

How much of the circle is shaded?

$\dfrac{3}{4}$ ← numerator / ← denominator

3 parts are shaded.
The circle is divided into 4 equal parts.

Read $\frac{3}{4}$ as three-fourths.

What fraction of the group of circles is shaded?

Four out of five circles are shaded.

$\dfrac{4}{5}$ ← circles shaded / ← circles in all

Read $\frac{4}{5}$ as four-fifths.

Practice

In problems 1–5, write the fraction *and* the word name to show how much of each figure or group is shaded.

1.

_____ or _____

2.

_____ or _____

3.

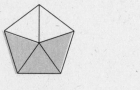

_____ or _____

4.

_____ or _____

5.

_____ or _____

6. Shade $\frac{2}{3}$ of this circle:

7. Shade $\frac{3}{5}$ of the group of squares below:

☐ ☐ ☐ ☐ ☐

8. Shade $\frac{7}{8}$ of the distance between 0 and 1:

0   1

9. What fraction of a dollar is shown below? _____

Picturing Equal Fractions

There is more than one way to write a proper fraction to represent a given amount.

For example, the two cups contain equal amounts of syrup.
• The cup at left is divided into 4 equal measuring units.
• The cup at right is divided into 8 equal measuring units.

$\frac{3}{4}$ and $\frac{6}{8}$ represent the same amount and are called **equal fractions** (or *equivalent fractions*).

Equal Amounts

$\frac{3}{4}$ full = $\frac{6}{8}$ full

Equal Fractions

Practice

■ **Write equal fractions for each pair of figures as indicated.**

1. fraction of each rectangle that's shaded

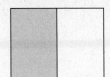 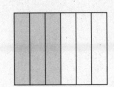

$$\frac{1}{2} = \frac{}{6}$$

2. fraction of each circle that's shaded

$$\frac{1}{3} = \frac{}{}$$

3. fraction of each pizza eaten

$$\frac{}{} = \frac{}{}$$

4. fraction of each group that's shaded

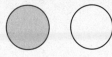

$$\frac{}{} = \frac{}{}$$

5. fraction of each cup filled with water

$$\frac{}{} = \frac{}{}$$

6. fraction of an inch that each bar measures

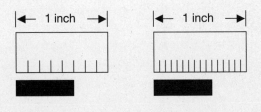

$$\frac{}{} = \frac{}{}$$

Reducing Fractions

To **reduce (simplify)** a fraction is to rewrite it as an equal fraction with smaller numbers.

When a fraction is in its simplest form—smallest numbers possible—it is said to be **reduced to lowest terms.**

Examples: $\frac{2}{4} = \frac{1}{2}$

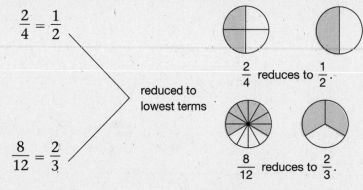

reduced to
lowest terms

$\frac{2}{4}$ reduces to $\frac{1}{2}$.

$\frac{8}{12} = \frac{2}{3}$

$\frac{8}{12}$ reduces to $\frac{2}{3}$.

To reduce a fraction to lowest terms, *divide* both numerator and denominator by the largest whole number that divides evenly into both.

Reduce $\frac{6}{9}$ to lowest terms.	Reduce $\frac{20}{32}$ to lowest terms.
Divide both numerator (6) and denominator (9) by 3. $\frac{6 \div 3}{9 \div 3} = \frac{2}{3}$	Divide both 20 and 32 by 2. $\frac{20 \div 2}{32 \div 2} = \frac{10}{16}$
	Divide again by 2.* $\frac{10 \div 2}{16 \div 2} = \frac{5}{8}$

*Sometimes, you just "don't see" the largest number you can divide by. When this happens, you may need to divide two or more times.

Practice

▦ **Reduce each fraction to lowest terms.**

1. $\frac{3}{9} =$ $\frac{2}{8} =$ $\frac{4}{6} =$ $\frac{6}{9} =$ $\frac{6}{8} =$

2. $\frac{4}{10} =$ $\frac{8}{12} =$ $\frac{9}{15} =$ $\frac{10}{14} =$ $\frac{12}{16} =$

3. $\frac{15}{45} =$ $\frac{14}{16} =$ $\frac{8}{32} =$ $\frac{10}{24} =$ $\frac{28}{64} =$

Writing an Amount as a Fraction of a Larger Unit

Measurement problems often require you to write one amount as a fraction of a larger unit.

What fraction of a yard is 32 inches? (1 yard = 36 inches)	What fraction of an hour is 40 minutes? (1 hour = 60 minutes)
Write: $\dfrac{32}{36} \leftarrow \dfrac{\text{part}}{\text{whole}}$	Write: $\dfrac{40}{60} \leftarrow \dfrac{\text{part}}{\text{whole}}$
Reduce: $\dfrac{32}{36} = \dfrac{32 \div 4}{36 \div 4} = \dfrac{8}{9}$	Reduce: $\dfrac{40}{60} = \dfrac{40 \div 20}{60 \div 20} = \dfrac{2}{3}$
Answer: 32 inches is $\frac{8}{9}$ of a yard.	Answer: 40 minutes is $\frac{2}{3}$ of an hour.

Practice

■ **Reduce each fraction to lowest terms.**

1. **1 yard = 36 inches**

   a) 12 inches = _____ yard

   b) 24 inches = _____ yard

   c) 27 inches = _____ yard

2. **1 meter = 100 centimeters**

   a) 20 centimeters = _____ meter

   b) 50 centimeters = _____ meter

   c) 60 centimeters = _____ meter

3. **1 pound = 16 ounces**

   a) 6 ounces = _____ pound

   b) 10 ounces = _____ pound

   c) 12 ounces = _____ pound

4. **1 hour = 60 minutes**

   a) 15 minutes = _____ hour

   b) 20 minutes = _____ hour

   c) 45 minutes = _____ hour

5. **1 year = 52 weeks**

   a) 4 weeks = _____ year

   b) 20 weeks = _____ year

   c) 32 weeks = _____ year

6. **1 cup = 8 fluid ounces**

   a) 4 fluid ounces = _____ cup

   b) 6 fluid ounces = _____ cup

   c) 8 fluid ounces = _____ cup

■ **Solve the problems.**

7. Ellen pays $400 each month for rent. If Ellen's monthly salary is $1,400, what fraction of her salary is needed to pay rent?

8. Miguel has delivered 12,000 pounds of an ordered 64,000 pounds of rock to the construction site. What fraction of the total needed has he delivered?

Raising Fractions to Higher Terms

Sometimes, instead of reducing a fraction, you want to rewrite it as an equal fraction with larger numbers. This is called **raising a fraction to higher terms**.

You often raise fractions to higher terms when you compare fractions and when you add or subtract fractions.

To raise a fraction to higher terms, *multiply* both numerator and denominator by the same number.

Equal Fractions

$$\frac{1}{2} = \frac{1 \times 2}{2 \times 2} = \frac{2}{4}$$

Write $\frac{3}{4}$ as a fraction that has 16 as a denominator.

Step 1. Look at the denominators.
Ask, "What do I multiply 4 by to get 16?"

Answer: 4

To Solve

$$\frac{3}{4} = \frac{?}{16}$$

Step 2. Multiply the numerator (3) by 4: $3 \times 4 = 12$
The new numerator is 12.

Answer: $\frac{3}{4} = \frac{12}{16}$

Think

$$\frac{3 \times 4}{4 \times 4} = \frac{12}{16}$$

Practice

Raise each fraction to higher terms.

1. $\overset{\times 4}{\underset{\times 4}{\frac{1}{2}}} = \frac{}{8}$ $\frac{1}{3} = \frac{}{6}$ $\frac{1}{4} = \frac{}{8}$ $\frac{2}{3} = \frac{}{9}$ $\frac{3}{4} = \frac{}{12}$

2. $\frac{2}{5} = \frac{}{15}$ $\frac{3}{4} = \frac{}{8}$ $\frac{2}{3} = \frac{}{12}$ $\frac{5}{7} = \frac{}{14}$ $\frac{2}{6} = \frac{}{18}$

3. $\overset{\times 5}{\underset{\times 5}{\frac{2}{3}}} = \frac{}{15}$ $\frac{2}{5} = \frac{}{30}$ $\frac{5}{14} = \frac{}{28}$ $\frac{11}{12} = \frac{}{36}$ $\frac{4}{7} = \frac{}{28}$

4. $\frac{5}{7} = \frac{}{42}$ $\frac{5}{6} = \frac{}{24}$ $\frac{7}{8} = \frac{}{24}$ $\frac{3}{4} = \frac{}{36}$ $\frac{5}{8} = \frac{}{24}$

Comparing Proper Fractions

Like fractions are fractions that have the same denominator—called a **common denominator.**

To compare like fractions, compare the numerators. The fraction with the larger numerator is the larger fraction, and so on.

Unlike fractions have different denominators. To compare unlike fractions, you rewrite them as like fractions.
- First, choose a common denominator. Often, the larger denominator of the compared fractions can be chosen.
- Next, rewrite each fraction as an equal fraction that has this common denominator.
- Finally, compare numerators of the like fractions.

Comparing **Like** Fractions

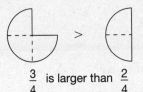

$\frac{3}{4}$ is larger than $\frac{2}{4}$

Comparing **Unlike** Fractions

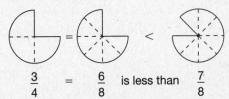

$\frac{3}{4} = \frac{6}{8}$ is less than $\frac{7}{8}$

Like Fractions	Unlike Fractions
Compare $\frac{3}{4}$ and $\frac{2}{4}$. $\frac{3}{4} > \frac{2}{4}$ because $3 > 2$.	Compare $\frac{3}{4}$ and $\frac{7}{8}$. *Step 1.* Choose 8 as the common denominator. *Step 2.* Write $\frac{3}{4}$ as an equal fraction having a denominator of 8. Rewrite $\frac{3}{4}$ as $\frac{?}{8}$: $\frac{3}{4} = \frac{6}{8}$. *Step 3.* Compare $\frac{6}{8}$ and $\frac{7}{8}$. $\frac{6}{8} < \frac{7}{8}$ because $6 < 7$. So, $\frac{3}{4} < \frac{7}{8}$.

Practice

Write >, <, or = to compare each pair of fractions below.

1. $\frac{3}{6}$ $\frac{2}{6}$ $\frac{5}{8}$ $\frac{7}{8}$ $\frac{15}{16}$ $\frac{13}{16}$ $\frac{19}{32}$ $\frac{23}{32}$

Rewrite unlike fractions as like fractions before comparing. Use the larger denominator in each pair as the common denominator.

2. $\frac{2}{3}$ $\frac{7}{12}$ $\frac{1}{4}$ $\frac{3}{12}$ $\frac{2}{5}$ $\frac{3}{10}$ $\frac{2}{3}$ $\frac{5}{6}$

3. $\frac{3}{4}$ $\frac{7}{8}$ $\frac{2}{3}$ $\frac{8}{9}$ $\frac{4}{7}$ $\frac{9}{14}$ $\frac{2}{5}$ $\frac{6}{15}$

Finding the Lowest Common Denominator (LCD)

To compare *unlike* fractions, you rewrite them as like fractions. The key step is choosing a *common denominator*.

What is a common denominator for $\frac{2}{3}$ and $\frac{3}{4}$?

The larger denominator (4) cannot be used: $\frac{2}{3} = \frac{?}{4}$. (Doesn't work!)

A common denominator is a number that each denominator divides into evenly. The smallest possible common denominator is called the **lowest common denominator** (LCD).

To find the lowest common denominator for $\frac{2}{3}$ and $\frac{3}{4}$ (and for other problems involving unlike fractions), we'll use the **method of comparing multiples**.

Compare the fractions $\frac{2}{3}$ and $\frac{3}{4}$.

Step 1. Write several multiples of each denominator.

Multiples of 3: 3   6   9   ⟨12⟩   15

Multiples of 4: 4   8   ⟨12⟩   16   20

Step 2. Choose the smallest number that is on both lists: 12

12 is the LCD.

Step 3. Rewrite each fraction as an equal fraction with a denominator of 12.

$\frac{2}{3} = \frac{8}{12}$ and $\frac{3}{4} = \frac{9}{12}$

Step 4. Compare the like fractions.

$\frac{8}{12} < \frac{9}{12}$, so $\frac{2}{3} < \frac{3}{4}$

Practice

▪ **Write five multiples of each of the following numbers.**

1. 2: 2, 4, 6, 8, 10   3:   4:   5:

2. 6:   8:   10:   12:

▪ **Use the method of comparing multiples to find the LCD for each pair of fractions.**

3. $\frac{1}{4}$ and $\frac{1}{3}$ $\frac{2}{3}$ and $\frac{1}{5}$ $\frac{1}{2}$ and $\frac{1}{3}$ $\frac{3}{8}$ and $\frac{1}{6}$

    4: 4, 8, ⟨12,⟩ 16   3:   2:   8:

    3: 3, 6, 9, ⟨12⟩   5:   3:   6:

    LCD: ____   LCD: ____   LCD: ____   LCD: ____

Math Tip

Another way to find a common denominator is to **multiply denominators by one another.**

Although this method always gives you a common denominator you can use, it does **not** always give you the lowest common denominator.

At right, both 48 and 24 can be used as common denominators, but 24 is smaller and easier to work with.

Find a common denominator for $\frac{1}{6}$ and $\frac{1}{8}$.

By Multiplying Denominators	By Comparing Multiples
$\frac{1}{6} = \frac{8}{48}$	$\frac{1}{6} = \frac{4}{24}$
$\frac{1}{8} = \frac{6}{48}$	$\frac{1}{8} = \frac{3}{24}$

■ **Write $>$, $<$, or $=$ to compare each pair of fractions below.**

Step 1. Determine the LCD. (If the largest denominator isn't the LCD, compare multiples to find it.)
Step 2. Use the LCD to rewrite each fraction.
Step 3. Compare the like fractions.

4. $\frac{2}{3}$ $\frac{4}{6}$ $\frac{1}{4}$ $\frac{1}{3}$ $\frac{2}{3}$ $\frac{2}{5}$ $\frac{2}{3}$ $\frac{3}{4}$

5. $\frac{1}{2}$ $\frac{3}{5}$ $\frac{7}{8}$ $\frac{3}{4}$ $\frac{2}{3}$ $\frac{1}{2}$ $\frac{2}{3}$ $\frac{5}{7}$

■ **Arrange each group of three fractions in order, writing the smallest fraction in each group first.**

Step 1. Use the largest denominator in each group as the LCD.
Step 2. Use the LCD to rewrite each fraction.
Step 3. Compare like fractions.

6. $\frac{2}{3}$ $\frac{7}{12}$ $\frac{3}{4}$ $\frac{2}{3}$ $\frac{7}{9}$ $\frac{13}{18}$ $\frac{3}{8}$ $\frac{1}{3}$ $\frac{7}{24}$ $\frac{6}{14}$ $\frac{4}{7}$ $\frac{1}{2}$

Writing Improper Fractions and Mixed Numbers

In an **improper fraction**, the numerator is either equal to or larger than the denominator.

$\frac{4}{4}$ The top 4 stands for the number of pieces you're referring to. The bottom 4 stands for the number of pieces in 1 whole.

Thus, the value of $\frac{4}{4}$ is 1.

$\frac{7}{4}$ One whole is divided into 4 pieces. But you're referring to 7 pieces. This is 3 pieces more than 1 whole.

Thus, the value of $\frac{7}{4}$ is larger than 1.

$\frac{4}{4}$

$\frac{7}{4}$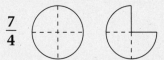

A **mixed number** is a whole number together with a fraction.

$1\frac{3}{4}$ The whole number *1* stands for 1 whole object.

Along with the 1 object, there is $\frac{3}{4}$ object more.

$1\frac{3}{4}$

Notice that $\frac{7}{4}$ and $1\frac{3}{4}$ represent the same amount. An amount larger than 1 can be written as either an improper fraction or a mixed number.

Practice

■ **Write either an improper fraction or a mixed number to represent each amount below.**

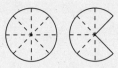

1. _____

3. _____

5. _____

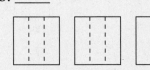

2. _____

4. _____

6. _____

■ **In problems 7–8, write your answer as an improper fraction and as a mixed number.**

7. While doing inventory, Martin counted 9 quarter-pound cubes of butter. How much butter is this?

   improper fraction: _____ pounds

   mixed number: _____ pounds

8. Glenna cuts pies into 8 equal pieces and then sells them by the slice. How many pies did Glenna sell if she sold 19 slices?

   improper fraction: _____ pies

   mixed number: _____ pies

Life Skill: Reading an English Ruler

Pictured below is the familiar 6-inch English ruler.

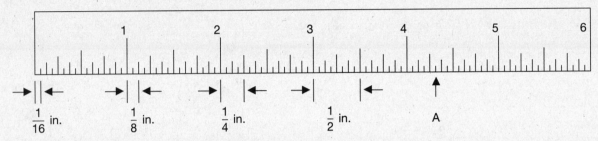

To make reading easier, each fraction of an inch is represented by a line of different height. For example, the smallest markings represent $\frac{1}{16}$ of an inch.

How far is point A from the left end of the ruler?

First notice that point A is between 4 and 5 inches from the left end.

Because point A is at a $\frac{1}{16}$ " line, count how many sixteenths point A is beyond 4.

Answer: $4\frac{5}{16}$ inches *or* $4\frac{5}{16}$ " (The symbol " stands for inches.)

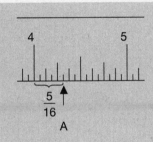

Practice

1. Write each of the ruler divisions as a number of sixteenths of an inch.

    a) $\frac{1}{8}$ in. = $\overline{}$ in.   b) $\frac{1}{4}$ in. = $\overline{16}$ in.   c) $\frac{1}{2}$ in. = $\overline{16}$ in.   d) 1 in. = $\overline{16}$ in.

2. What distance is represented in each ruler pictured below? Reduce fraction answers.

    a)    b) 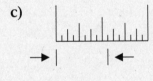   c) 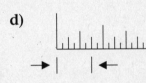   d)

    _____ inch   _____ inch   _____ inch   _____ inch

3.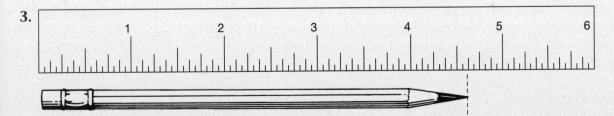

    What is the length of the pencil to the nearest

    a) $\frac{1}{16}$ " = _____   b) $\frac{1}{8}$ " = _____   c) $\frac{1}{4}$ " = _____   d) $\frac{1}{2}$ " = _____

Understanding Percent

Percent is the third way to write part of the whole.

Percent means *parts out of 100.*

 is 1% of

You write percent as the *number of hundredths* followed by the percent sign (%).
- A *5% sales tax* means that you must pay $0.05 tax for every dollar of your purchase price.
- A *4% savings rate* means that $100 in savings will earn $4 in interest each year.

Dividing a whole into 100 equal parts is an easy way to visualize percent.

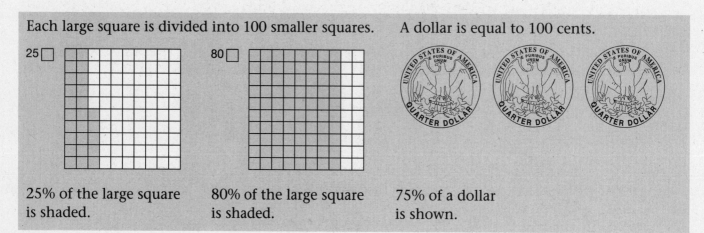

Each large square is divided into 100 smaller squares.

25☐

80☐

A dollar is equal to 100 cents.

25% of the large square is shaded.

80% of the large square is shaded.

75% of a dollar is shown.

What is 100%?

100% stands for a whole amount. $\boxed{100\% = 1}$

At right, 35% of the large square is shaded; 65% is unshaded.

35% + 65% = 100%

shaded + unshaded = whole square

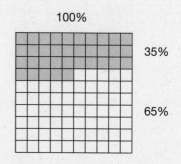

100%

35%

65%

How About Percents Larger than 100%?

A percent larger than 100% represents a number larger than 1.

200% stands for *2 whole objects,* or *twice an amount.*
300% stands for *3 whole objects,* or *3 times an amount,* and so on.

Percents larger than 100% are not as common, but you need to understand them.

Math Tip

A sales increase from $100 to $300 is a **200% increase.**

Amount Increase
$300 − $100 = $200

Percent Increase
200% because $200 is *2 × $100.*

Practice

▇ **In problems 1–4, write what percent of each square below is shaded, what percent is unshaded, and what the total of the two percents equals.**

1.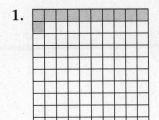

   % shaded _____

   % unshaded _____

   total % _____

3.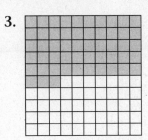

   % shaded _____

   % unshaded _____

   total % _____

2.

   % shaded _____

   % unshaded _____

   total % _____

4.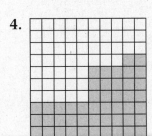

   % shaded _____

   % unshaded _____

   total % _____

▇ **Solve each problem.**

5. Chase County is trying to raise $100,000 in this year's United Way Fund Drive.

   a) If the fund-raisers hope to raise 40% of the total during the first 2 weeks of the drive, how much will they need to collect during these first 14 days?

   b) After collecting $85,000, what percent of the goal will they have reached?

6. The Kingston Classic car race pits some of the country's best drivers against each other for 100 laps around a 2-mile-long track.

   a) What percent of the race does Mario have remaining after he's completed 42 laps?

   b) After driving 150 miles, what percent of the race has Mario finished? (Hint: $\frac{150}{200} = \frac{?}{100}$)

7. About 100,000 people attended last year's Rose Bowl game. The announcer said, "Had seats been available, we could have sold 400,000 tickets!"

   a) If the announcer was right, how many additional tickets could have been sold?

   b) What percent increase in sales do these additional tickets represent?

8. Two years ago, Brett Furniture had sales of $500,000. Last year it had sales of $2,500,000.

   a) By how much money did Brett's sales increase from one year to the next? Express your answer in dollars.

   b) By what percent did Brett's sales increase from one year to the next?

Relating Decimals, Fractions, and Percents

A percent can easily be written as an equivalent decimal or fraction.
- Percent has the same value as a two-place decimal.
  37% is equal to **0.37.**
- Percent has the same value as a fraction that has a denominator of 100.
  37% is equal to $\frac{37}{100}$.

100 Equal Parts

37% is shaded.

0.37 is shaded.

$\frac{37}{100}$ is shaded.

Math Tip

To remember the meaning of percent, many students just look at the % sign.

- To write a percent as a decimal, let the two zeros in % remind you of two decimal places: 37% = 0.37.

- To write the percent as a fraction, let the two zeros in percent remind you of two zeros in the denominator: $\frac{37}{100}$.

Practice

Change each percent to an equivalent decimal.

1. 25%   37%   50%   78%   98%

Write each percent as an equivalent whole number.

2. 300%   600%   200%   400%   500%

Write each percent as an equivalent fraction. Reduce each fraction when possible.

3. 20%   25%   50%   60%   75%

Express each percent shown below as a decimal (d) and as a fraction (f).

4. **Save 20%**   5. **Savings Rate**   6. **New Home Loans**
                            Now 5%   **30-Year Fixed Rate 12%**

d: _____ f: _____   d: _____ f: _____   d: _____ f: _____

Determine whether the following number sentences are true or false. Circle T if a statement is true or F if it is false.

7. $25\% < \frac{1}{2}$ T or F $50\% > \frac{3}{4}$ T or F $20\% = \frac{2}{5}$ T or F

8. $40\% < \frac{1}{3}$ T or F $70\% \ne .07$ T or F $200\% > .250$ T or F

Answer each question. Reduce fractions to lowest terms.

9. Brent rode in a 100-mile bike race.

   a) What percent of the race had he finished after riding 60 miles?

   b) At this point, what *fraction* of the race did he still have left to ride?

10. By 9:30, Aaron had completed $\frac{3}{5}$ of his homework.

   a) By 9:30, what percent of his homework had Aaron completed?
       (Hint: $\frac{3}{5} = \frac{?}{100}$)

   b) How do you write $\frac{3}{5}$ as a decimal?

11. A **meter** (a little longer than a yard) is a metric length unit that is equal to 100 *centimeters*.

   a) Expressed as a fraction, what part of a meter is 90 centimeters (about 1 yard)?

   b) Expressed as a percent, what part of a meter is 30 centimeters (about 1 foot)?

   c) Expressed as a decimal, what part of a meter is a length of 63 centimeters?

12. A ton is equal to 2,000 pounds.

   a) What fraction of a ton is 1,500 pounds?

   b) What percent of a ton is 1,500 pounds?
       (Hint: $\frac{1,500}{2,000} = \frac{?}{100}$)

13. Julia went to a sale where "all blouses are marked down 25%."

   a) Express this price reduction as a decimal.

   b) Express this price reduction as a fraction.

14. A **kilometer** (about 0.6 mile) is a metric length unit that is equal to 1,000 meters. What part of a kilometer is 450 meters? Express your answer in three ways:

   a) as a decimal: _____
       (Hint: 1 meter = 0.001 kilometer.)

   b) as a fraction reduced to hundredths: _____
       (Hint: $\frac{450}{1,000} = \frac{?}{100}$.)

   c) as a percent: _____
       (Hint: See answer b.)

Data Highlight: Reading a Circle Graph

The most common way to graphically display parts of a whole is by using a circle graph.
- An entire circle represents a whole amount.
- Individual segments represent parts of the whole, the largest segment representing the largest portion, and so on.

On a circle graph, the sum of all segments must add up to 1 or 100%.

Math Tip

Because it resembles a cut pie, a circle graph is often called a **pie graph** or **pie chart**.

The Monroe family budget of $15,000 is divided into 6 expense categories. Using a circle graph, you can display the Monroe budget in any of 3 ways.

A. Decimals*
(Each expense is displayed as *cents per budget dollar*.)

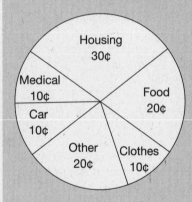

B. Fractions
(Each expense is displayed as a *fraction of the total*.)

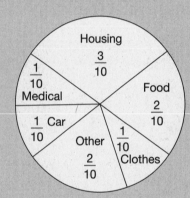

C. Percents
(Each expense is displayed as a *percent of the total*.)

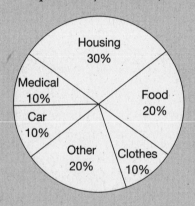

*In a circle graph, an amount such as thirty cents can be written as 30¢ or as $0.30.

Graph A: How many more cents per dollar do the Monroes spend on food than on clothes?
Answer: 20¢ − 10¢ = 10¢

Graph B: Which is the largest expense category of the Monroe family?
Answer: Housing ($\frac{3}{10}$ is the largest fraction shown.)

Graph C: What total percent of the Monroe family budget is spent on medical, food, *and* housing?
Answer: 10% + 20% + 30% = 60%

Practice

■ **Problems 1–3 refer to Graph A.**

Graph A
QRT Boat Company Sales

¢ per $ of sales

1. How many more cents per dollar of sales does the QRT Boat Company make from North American sales than from European sales?

2. What number of cents should be written in the segment labeled "Other"?

3. Are Pacific Rim sales double the amount of European sales?

■ **Problems 4–6 refer to Graph B.**

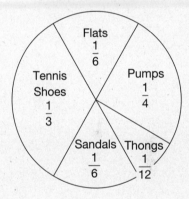

Graph B
Women's Shoes Preference Poll

4. Write the fractions indicated on the preference poll as equal fractions all having a common denominator.

   Flats: $\frac{1}{6}$ =   Pumps: $\frac{1}{4}$ =   Sandals: $\frac{1}{6}$ =

   Thongs: $\frac{1}{12}$ =   Tennis Shoes: $\frac{1}{3}$ =

5. According to the results of the preference poll, which type of shoe is most preferred by the women who were surveyed?

6. Which fraction is the *mode* of the results found in the preference poll?

■ **Problems 7–9 refer to Graph C.**

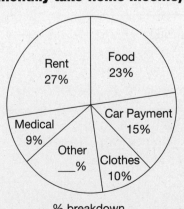

Graph C
Smith Family Monthly Budget
(monthly take-home income)

% breakdown

7. What total percent of their monthly take-home income do the Smiths spend on rent, food, *and* car payment?

8. Of the specific expense categories listed (not counting "Other"), which of the five percents is the *median?*

9. What percent of the Smith's family take-home income should be written in the segment labeled "Other"?

Test Readiness Checkup

Circle each correct answer. Check your answers on page 242; then correct any errors.

1. A calculator displayed the following answer to a subtraction problem.

   0.002

   In words, this number is read as:

   (1) two ten-thousandths
   (2) two-thousandths
   (3) two-hundredths
   (4) two-tenths
   (5) two

2. Kim was told that the hole she was to drill for the shaft should be "about thirty-two hundredths of an inch."

   This number is written in digits as:

   (1) 32   (4) 0.032
   (2) 3.2   (5) 0.0032
   (3) 0.32

3. Which of the following inequalities is *not* true?

   (1) $0.8 > 0.79$ (4) $2.5 < 1.99$
   (2) $0.14 < 0.23$ (5) $1.4 > 0.75$
   (3) $1.3 > 1.25$

4. In a 100-meter dash, the following times were recorded:

   Al: 11.34 sec.   Jess: 10.82 sec.
   Bill: 10.7 sec.   Bobby: 11.09 sec.

   Listing the winner first, what is the correct finishing order of these racers?

   (1) Bill, Jess, Bobby, Al
   (2) Jess, Bobby, Bill, Al
   (3) Al, Bobby, Jess, Bill
   (4) Bobby, Jess, Al, Bill
   (5) Jess, Bill, Bobby, Al

Problem 5 refers to the drawing below.

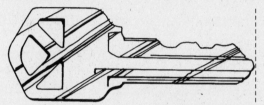

5. Students in Mr. Alvarez's class measured the key as shown above. The students gave these four answers:

   1. 68 mm   3. 6.08 cm
   2. 608 mm   4. 6.8 cm

   Which of these answers is correct?

   (1) 1 only   (4) 2 only
   (2) 1 and 3   (5) 2 and 4
   (3) 1 and 4

6. Forty minutes can be written as a fraction of an hour. How is this fraction written in its most reduced form?

   (1) $\frac{40}{60}$ (4) $\frac{2}{3}$

   (2) $\frac{4}{6}$ (5) $\frac{3}{2}$

   (3) $\frac{60}{40}$

7. What is the lowest common denominator for the fractions $\frac{3}{4}$ and $\frac{5}{6}$?

   (1) 4
   (2) 6
   (3) 10
   (4) 12
   (5) 24

8. Judy has been asked to list the following three board thicknesses in order of size:

$\frac{7}{12}$ inch $\frac{2}{3}$ inch $\frac{3}{4}$ inch
Board A   Board B   Board C

Listing the *narrowest board first,* what is the correct order of sizes?

(1) A, C, B   (4) B, C, A
(2) A, B, C   (5) C, A, B
(3) B, A, C

9. A number larger than 1 can be written as a mixed number *or* as an improper fraction.

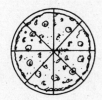

Which of the following does *not* correctly represent the amount of pizza shown above?

(1) $\frac{12}{8}$ (4) $1\frac{1}{2}$

(2) $1\frac{4}{8}$ (5) $\frac{6}{4}$

(3) $\frac{8}{12}$

■ **Problem 10 refers to the drawing below.**

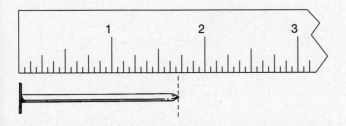

10. To the nearest $\frac{1}{16}$ inch, what is the length in inches of the nail shown above?

(1) $1\frac{15}{16}$ (4) $1\frac{3}{4}$

(2) $1\frac{5}{8}$ (5) $1\frac{13}{16}$

(3) $1\frac{11}{16}$

11. Beth baked 100 pastries for the school picnic. Of these 100 pastries, 65 were doughnuts and the rest were croissants.

What percent of Beth's pastries were croissants?

(1) 25%   (4) 100%
(2) 35%   (5) 165%
(3) 65%

12. A **kilogram** (about 2.2 pounds) is a metric unit of weight that's equal to 1,000 grams. Expressed as a percent, what part of a kilogram is a weight of 250 grams?

(1) 25%   (4) 75%
(2) 45%   (5) 95%
(3) 55%

■ **Problem 13 refers to the graph below.**

Thompson Shoe Sales (by country)

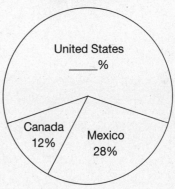

13. What percent of sales of the Thompson Shoe Company is made in the United States?

(1) 12%   (4) 60%
(2) 28%   (5) 88%
(3) 40%

UNIT 4 Fractions

Building Confidence by Estimating

People often feel unsure of their answers when working with fractions. Why is this? Here's a typical response:

"I just don't have a feeling about how answers should turn out."

If you've had a similar feeling, estimation can help. You can use estimation to find answers, check answers, and help choose among answer choices on a test.

As a first step, learn to recognize when a fraction is close to 0, $\frac{1}{2}$, or 1.

- A fraction is close to 1 when the denominator and numerator are about the same size.

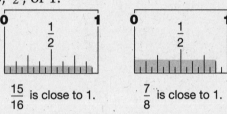

$\frac{15}{16}$ is close to 1. $\frac{7}{8}$ is close to 1.

- A fraction is close to $\frac{1}{2}$ when the denominator is about twice as large as the numerator.

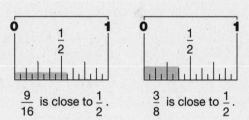

$\frac{9}{16}$ is close to $\frac{1}{2}$. $\frac{3}{8}$ is close to $\frac{1}{2}$.

- A fraction is close to 0 when the numerator is very small compared to the denominator.

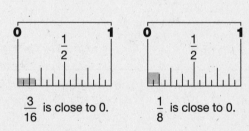

$\frac{3}{16}$ is close to 0. $\frac{1}{8}$ is close to 0.

Estimating to Add Proper Fractions

Estimating is especially useful when you are adding proper fractions. The first step is to round each fraction.

- Round a fraction to 1 if the denominator and numerator are about equal.
  Fractions that round to 1: $\frac{3}{4}, \frac{5}{6}, \frac{7}{8}, \frac{9}{10}, \frac{11}{12}, \frac{15}{16}$

- Round a fraction to $\frac{1}{2}$ if the denominator is about twice the numerator.
  Fractions that round to $\frac{1}{2}$: $\frac{1}{3}, \frac{2}{3}, \frac{3}{5}, \frac{3}{8}, \frac{5}{9}, \frac{7}{12}$

- Round a fraction to 0 if the numerator is much smaller than the denominator.
  Fractions that round to 0: $\frac{1}{5}, \frac{1}{6}, \frac{2}{7}, \frac{1}{8}, \frac{1}{10}, \frac{1}{12}, \frac{3}{16}$

Math Tip

A fraction that's close to two different values can be rounded to either one! Either estimate will work fine. Remember to allow yourself flexibility when estimating.

When you are estimating a sum of proper fractions, group two halves ($\frac{1}{2}$) to make a whole (1). Then add the 1s, $\frac{1}{2}$s, and 0s.

Estimate the following sums.

$$\frac{3}{8} \to \frac{1}{2}$$
$$\frac{4}{9} \to \frac{1}{2} \quad \Big\} \; 1$$
$$+ \; \frac{6}{7} \to 1$$
Estimate: 2

$$\frac{4}{7} \to \frac{1}{2}$$
$$\frac{3}{6} \to \frac{1}{2} \quad \Big\} \; 1$$
$$+ \; \frac{1}{8} \to 0$$
Estimate: 1

$$\frac{11}{12} \to 1$$
$$\frac{7}{8} \to 1$$
$$\frac{9}{10} \to 1$$
$$+ \; \frac{1}{2} \to \frac{1}{2}$$
Estimate: $3\frac{1}{2}$

$$\frac{15}{16} \to 1$$
$$\frac{1}{16} \to 0$$
$$\frac{3}{8} \to \frac{1}{2}$$
$$+ \; \frac{5}{6} \to 1$$
Estimate: $2\frac{1}{2}$

Practice

Estimate a sum for each problem below.

1.

$$\frac{2}{4}$$
$$\frac{7}{8}$$
$$+ \; \frac{1}{6}$$

$$\frac{7}{8}$$
$$\frac{1}{5}$$
$$+ \; \frac{2}{3}$$

$$\frac{8}{9}$$
$$\frac{3}{4}$$
$$+ \; \frac{2}{5}$$

$$\frac{5}{6}$$
$$\frac{1}{2}$$
$$+ \; \frac{9}{10}$$

$$\frac{15}{16}$$
$$\frac{3}{7}$$
$$+ \; \frac{1}{3}$$

Solve each problem below with a reasonable estimate.

2. The three spacers below are to be placed end to end. What will be their approximate combined length?

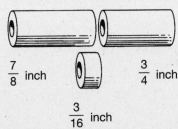

$\frac{7}{8}$ inch $\frac{3}{4}$ inch

$\frac{3}{16}$ inch

3. Every Monday, Wednesday, and Friday, Lydia swims $\frac{1}{2}$ mile. Tuesdays and Thursdays she swims $\frac{7}{8}$ mile. About how many miles does Lydia swim each week?

Estimating with Mixed Numbers

To estimate with mixed numbers, round each mixed number to the nearest whole number.
- If the value of the fraction is $\frac{1}{2}$ or more, round to the next higher whole number.
- If the value of the fraction is less than $\frac{1}{2}$, drop the fraction.

Example 1. Estimate $14\frac{11}{12} - 7\frac{1}{3}$.	**Example 2.** Estimate $9\frac{1}{4} \times 7\frac{2}{3}$.
Step 1. Round each mixed number to the nearest whole number.	*Step 1.* Round each mixed number to the nearest whole number.
$14\frac{11}{12} \to 15 \quad 7\frac{1}{3} \to 7$	$9\frac{1}{4} \to 9 \quad 7\frac{2}{3} \to 8$
$(\frac{11}{12} > \frac{1}{2}) \quad (\frac{1}{3} < \frac{1}{2})$	$(\frac{1}{4} < \frac{1}{2}) \quad (\frac{2}{3} > \frac{1}{2})$
Step 2. Subtract: $15 - 7 = 8$	*Step 2.* Multiply: $9 \times 8 = 72$
Answer: $14\frac{11}{12} - 7\frac{1}{3} \approx 8$	**Answer:** $9\frac{1}{4} \times 7\frac{2}{3} \approx 72$

Note: The $\approx$ symbol means "is approximately equal to."

Practice

Estimate an answer for each problem below.

4.
$$9\frac{2}{3} \atop +7\frac{1}{4}$$
$$7\frac{1}{8} \atop +3\frac{2}{3}$$
$$16\frac{7}{10} \atop +\ 9$$
$$24\frac{1}{2} \atop +17\frac{8}{9}$$
$$32\frac{11}{16} \atop +21\frac{1}{3}$$

5.
$$12\frac{1}{2} \atop -9\frac{7}{8}$$
$$9\frac{7}{8} \atop -3\frac{1}{4}$$
$$23\frac{15}{16} \atop -8\frac{1}{16}$$
$$18\frac{1}{10} \atop -12$$
$$11\frac{5}{12} \atop -6\frac{7}{8}$$

6. $8\frac{3}{4} \times 7\frac{1}{2} \approx$ $\qquad$ $5\frac{1}{3} \times 4\frac{4}{5} \approx$ $\qquad$ $14 \times 10\frac{1}{4} \approx$ $\qquad$ $20\frac{5}{8} \times 8\frac{1}{5} \approx$

To estimate when dividing, round to compatible whole numbers and then divide.

7. $14\frac{1}{2} \div 3\frac{7}{8} \approx$ $\qquad$ $23\frac{2}{3} \div 8\frac{1}{4} \approx$ $\qquad$ $73\frac{7}{8} \div 9\frac{1}{4} \approx$ $\qquad$ $65\frac{1}{3} \div 9\frac{1}{2} \approx$

Solve each problem below with a reasonable estimate.

8. While at the deli, Linda bought $1\frac{3}{4}$ pounds of sliced turkey, $2\frac{7}{8}$ pounds of sliced beef, and $4\frac{3}{16}$ pounds of sliced ham. About how much meat did Linda buy in all?

9. Linda's husband, Mario, bought $1\frac{7}{8}$ pounds of pastrami for $3.89 per pound and $2\frac{1}{8}$ pounds of sliced chicken for $2.19 per pound. Approximately how much did Mario's purchase cost?

10. During the weekend storm, $5\frac{1}{4}$ inches of snow fell on Saturday and $4\frac{7}{8}$ inches on Sunday. An additional $3\frac{1}{2}$ inches fell on Monday. About how much snow fell on these three days?

11. From a piece of oak molding that is $32\frac{1}{2}$ inches long, Sue plans to cut off two pieces: one measuring $14\frac{1}{4}$ inches and one $9\frac{7}{8}$ inches. Which expression below gives the best estimate of how many inches of molding Sue will have left after cutting off these two pieces?

   (1) $33 - 14 + 10$
   (2) $33 - (14 + 10)$
   (3) $32 - 14 - 9$
   (4) $(14 + 10) - 33$
   (5) $(14 - 10) + 33$

12. A one-cubic-foot container can hold $7\frac{1}{2}$ gallons of liquid. About how many gallons of gasoline does it take to fill a gas tank that has a volume of $3\frac{1}{8}$ cubic feet?

13. The value of TAQ stock slid last week from $56\frac{1}{4}$ to $38\frac{7}{8}$ following news of a major lawsuit against the company. By about how much did the value of TAQ stock drop last week?

14. Losing weight at the rate of $3\frac{1}{4}$ pounds per month, approximately how many months will it take Yoshi to lose a total of $28\frac{3}{4}$ pounds?

15. For Halloween, Timothy bought a $15\frac{1}{4}$-pound bag of candy chews at bulk rate. He took out $2\frac{1}{8}$ pounds for his family and then divided the rest into 50 "treat bags." Which expression gives the best estimate of how much candy will be in each treat bag?

   (1) $\frac{50}{15-2}$ lb.   (4) $\frac{15+2}{50}$ lb.

   (2) $\frac{50}{15+2}$ lb.   (5) $\frac{15-2}{50}$ lb.

   (3) $\frac{15-2}{50+17}$ lb.

Changing an Improper Fraction to a Mixed Number

Many answers to problems involving fractions are first written as improper fractions. To simplify these answers, you change them to mixed numbers.

To change an improper fraction to a mixed number
- divide the denominator into the numerator
- write the remainder as a fraction

Example 1. Change $\frac{14}{3}$ to a mixed number.	**Example 2.** Change $\frac{12}{8}$ to a mixed number.
Step 1. Divide 3 into 14:	*Step 1.* Divide 8 into 12:
$$\begin{array}{r} 4\ r\ 2 \\ 3\overline{)14} \\ -12 \\ \hline 2 \end{array}$$	$$\begin{array}{r} 1\ r\ 4 \\ 8\overline{)12} \\ -\ 8 \\ \hline 4 \end{array}$$
Step 2. Write the remainder (2) over the divisor (3) to form the fraction part of the answer.	*Step 2.* Write a fraction: $1\ r4 = 1\frac{4}{8}$ *Step 3.* Reduce the fraction. $\frac{4 \div 4}{8 \div 4} = \frac{1}{2}$
Answer: $4\frac{2}{3}$	Answer: $1\frac{4}{8} = 1\frac{1}{2}$

Practice

■ **Change each improper fraction to a mixed number.**

1. $\frac{7}{3} =$ $\frac{13}{10} =$ $\frac{9}{2} =$ $\frac{11}{3} =$ $\frac{14}{5} =$

2. $\frac{9}{6} =$ $\frac{18}{4} =$ $\frac{14}{4} =$ $\frac{23}{5} =$ $\frac{26}{6} =$

3. At his deli, Sal gives away samples of salami in $\frac{1}{12}$ -pound packages. How many pounds of salami will Sal need to make 42 sample packages? Express your answer as an improper fraction *and* as a mixed number.

   Improper fraction: _____

   Mixed number: _____

4. For the company party, Connie ordered 14 pizzas, each divided into 8 slices. At the end of the party, Connie gathered up 18 slices of uneaten pizza. How many pizzas were left over? Express your answer as an improper fraction *and* as a mixed number.

   Improper fraction: _____

   Mixed number: _____

Adding Like Fractions

To add like fractions, add the numerators and place the sum over the denominator. Reduce the answer to lowest terms.

Example 1. Add $\frac{13}{16}$ and $\frac{9}{16}$.

Step 1. Add the numerators: $13 + 9 = 22$

Step 2. Place the sum over 16.

$$\frac{13}{16} + \frac{9}{16} = \frac{22}{16}$$

Step 3. Change $\frac{22}{16}$ to a mixed number and reduce the fraction.

$$\frac{22}{16} = 1\frac{6}{16} = 1\frac{3}{8}$$

Answer: $1\frac{3}{8}$

Example 2. Add $3\frac{5}{8}$ and $2\frac{7}{8}$.

Step 1. Add the fractions.

Step 2. Add the whole numbers.

$$\begin{array}{r} 3\frac{5}{8} \\ + 2\frac{7}{8} \\ \hline 5\frac{12}{8} \end{array}$$

Step 3. Change $\frac{12}{8}$ to a mixed number.

$$5\frac{12}{8} = 5 + 1\frac{4}{8}\ (\text{since } \frac{12}{8} = 1\frac{4}{8})$$

Step 4. Add the 1 to the 5: $5 + 1\frac{4}{8} = 6\frac{4}{8}$

Reduce the fraction: $6\frac{4}{8} = 6\frac{1}{2}$

Answer: $6\frac{1}{2}$

Practice

■ **Add. Reduce each answer to lowest terms.**

Fractions That Add to Less than 1

1.
$$\begin{array}{r} \frac{1}{3} \\ + \frac{1}{3} \\ \hline \frac{2}{3} \end{array} \qquad \begin{array}{r} \frac{2}{4} \\ + \frac{1}{4} \\ \hline \end{array} \qquad \begin{array}{r} \frac{5}{8} \\ + \frac{2}{8} \\ \hline \end{array} \qquad \begin{array}{r} \frac{2}{5} \\ + \frac{2}{5} \\ \hline \end{array} \qquad \begin{array}{r} \frac{8}{16} \\ + \frac{7}{16} \\ \hline \end{array}$$

2. $\frac{3}{8} + \frac{1}{8} = \frac{4}{8}$ $\frac{2}{6} + \frac{1}{6} =$ $\frac{3}{8} + \frac{3}{8} =$ $\frac{1}{4} + \frac{1}{4} =$ $\frac{4}{12} + \frac{4}{12} =$

$$\frac{4 \div 4}{8 \div 4} = \frac{1}{2}$$

3.
$$\begin{array}{r} \frac{3}{8} \\ \frac{2}{8} \\ + \frac{1}{8} \\ \hline \end{array} \qquad \begin{array}{r} \frac{4}{9} \\ \frac{3}{9} \\ + \frac{1}{9} \\ \hline \end{array} \qquad \begin{array}{r} \frac{3}{10} \\ \frac{3}{10} \\ + \frac{2}{10} \\ \hline \end{array} \qquad \begin{array}{r} \frac{2}{6} \\ \frac{1}{6} \\ + \frac{1}{6} \\ \hline \end{array} \qquad \begin{array}{r} \frac{5}{12} \\ \frac{3}{12} \\ + \frac{2}{12} \\ \hline \end{array}$$

$$\frac{6}{8} = \frac{6 \div 2}{8 \div 2} = \frac{3}{4}$$

Fractions That Add to a Whole Number

4.

$$\begin{array}{r}\frac{4}{5}\\+\ \frac{1}{5}\\\hline\frac{5}{5}=1\end{array}\qquad\begin{array}{r}\frac{5}{8}\\+\ \frac{3}{8}\\\hline\end{array}\qquad\begin{array}{r}\frac{3}{4}\\+\ \frac{1}{4}\\\hline\end{array}\qquad\begin{array}{r}\frac{2}{3}\\+\ \frac{1}{3}\\\hline\end{array}\qquad\begin{array}{r}\frac{11}{16}\\+\ \frac{5}{16}\\\hline\end{array}$$

5. $\quad\frac{3}{4}+\frac{2}{4}+\frac{3}{4}=\frac{8}{4}=2\qquad\qquad\frac{2}{3}+\frac{2}{3}+\frac{2}{3}=\qquad\qquad\frac{7}{8}+\frac{6}{8}+\frac{5}{8}+\frac{6}{8}=$

Fractions That Add to More than 1

6.

$$\begin{array}{r}\frac{3}{4}\\\frac{2}{4}\\+\\\hline\frac{5}{4}=1\frac{1}{4}\end{array}\qquad\begin{array}{r}\frac{2}{3}\\\frac{2}{3}\\+\\\hline\end{array}\qquad\begin{array}{r}\frac{7}{8}\\\frac{6}{8}\\+\\\hline\end{array}\qquad\begin{array}{r}\frac{11}{12}\\\frac{6}{12}\\+\\\hline\end{array}\qquad\begin{array}{r}\frac{5}{6}\\\frac{2}{6}\\+\\\hline\end{array}$$

7.

$$\begin{array}{r}\frac{3}{4}\\\frac{3}{4}\\+\\\hline\frac{6}{4}=1\frac{2}{4}=1\frac{1}{2}\end{array}\qquad\begin{array}{r}\frac{6}{8}\\\frac{4}{8}\\+\\\hline\end{array}\qquad\begin{array}{r}\frac{5}{6}\\\frac{5}{6}\\+\\\hline\end{array}\qquad\begin{array}{r}\frac{9}{10}\\\frac{6}{10}\\+\\\hline\end{array}\qquad\begin{array}{r}\frac{8}{12}\\\frac{6}{12}\\+\\\hline\end{array}$$

Adding Fractions and Mixed Numbers

8.

$$\begin{array}{r}1\frac{3}{4}\\+\ \frac{3}{4}\\\hline 1\frac{6}{4}=1+1\frac{2}{4}\\=2\frac{2}{4}=2\frac{1}{2}\end{array}\qquad\begin{array}{r}2\frac{2}{3}\\+\ \frac{2}{3}\\\hline\end{array}\qquad\begin{array}{r}1\frac{3}{8}\\+\ \frac{7}{8}\\\hline\end{array}\qquad\begin{array}{r}5\frac{1}{2}\\+\ \frac{1}{2}\\\hline\end{array}\qquad\begin{array}{r}2\frac{9}{16}\\+\ \frac{11}{16}\\\hline\end{array}$$

9.

$$\begin{array}{r}3\frac{4}{6}\\+2\frac{5}{6}\\\hline 5\frac{9}{6}=5+1\frac{3}{6}\\=6\frac{3}{6}=6\frac{1}{2}\end{array}\qquad\begin{array}{r}2\frac{5}{8}\\+1\frac{7}{8}\\\hline\end{array}\qquad\begin{array}{r}4\frac{3}{5}\\+2\frac{2}{5}\\\hline\end{array}\qquad\begin{array}{r}5\frac{11}{12}\\+3\frac{9}{12}\\\hline\end{array}\qquad\begin{array}{r}8\frac{9}{16}\\+7\frac{5}{16}\\\hline\end{array}$$

Subtracting Like Fractions

To subtract like fractions, subtract the numerators and place the difference over the denominator. Reduce the answer to lowest terms.

Example 1. Subtract $\frac{3}{8}$ from $\frac{7}{8}$.

Step 1. Subtract the numerators.

Step 2. Place the difference (4) over the denominator

$$\begin{array}{r} \frac{7}{8} \\ -\frac{3}{8} \\ \hline \frac{4}{8} \end{array}$$

Step 3. Reduce the fraction $\frac{4}{8}$: $\frac{4 \div 4}{8 \div 4} = \frac{1}{2}$

Answer: $\frac{4}{8} = \frac{1}{2}$

Example 2. Subtract $2\frac{1}{4}$ from $5\frac{3}{4}$.

Step 1. Subtract the fractions.

Step 2. Subtract the whole numbers.

$$\begin{array}{r} 5\frac{3}{4} \\ -2\frac{1}{4} \\ \hline 3\frac{2}{4} \end{array}$$

Step 3. Reduce the fraction $\frac{2}{4}$: $\frac{2 \div 2}{4 \div 2} = \frac{1}{2}$

Answer: $3\frac{2}{4} = 3\frac{1}{2}$

Practice

Subtract. Reduce each answer to lowest terms.

1.
$\begin{array}{r} \frac{3}{4} \\ -\frac{2}{4} \\ \hline \end{array}$
$\qquad$
$\begin{array}{r} \frac{6}{8} \\ -\frac{3}{8} \\ \hline \end{array}$
$\qquad$
$\begin{array}{r} \frac{4}{5} \\ -\frac{2}{5} \\ \hline \end{array}$
$\qquad$
$\begin{array}{r} \frac{7}{9} \\ -\frac{5}{9} \\ \hline \end{array}$
$\qquad$
$\begin{array}{r} \frac{9}{10} \\ -\frac{6}{10} \\ \hline \end{array}$

2. $\quad \frac{6}{8} - \frac{2}{8} = \qquad \frac{8}{9} - \frac{2}{9} = \qquad \frac{11}{12} - \frac{5}{12} = \qquad \frac{13}{16} - \frac{9}{16} =$

Subtracting Fractions from Mixed Numbers

3.
$\begin{array}{r} 2\frac{7}{8} \\ -\frac{3}{8} \\ \hline \end{array}$
$\qquad$
$\begin{array}{r} 1\frac{3}{4} \\ -\frac{1}{4} \\ \hline \end{array}$
$\qquad$
$\begin{array}{r} 3\frac{2}{3} \\ -\frac{1}{3} \\ \hline \end{array}$
$\qquad$
$\begin{array}{r} 1\frac{9}{10} \\ -\frac{5}{10} \\ \hline \end{array}$
$\qquad$
$\begin{array}{r} 2\frac{11}{12} \\ -\frac{3}{12} \\ \hline \end{array}$

Subtracting Mixed Numbers

4.
$\begin{array}{r} 3\frac{3}{4} \\ -2\frac{1}{4} \\ \hline \end{array}$
$\qquad$
$\begin{array}{r} 4\frac{7}{8} \\ -2\frac{1}{8} \\ \hline \end{array}$
$\qquad$
$\begin{array}{r} 6\frac{5}{6} \\ -3\frac{3}{6} \\ \hline \end{array}$
$\qquad$
$\begin{array}{r} 12\frac{13}{16} \\ -4\frac{3}{16} \\ \hline \end{array}$
$\qquad$
$\begin{array}{r} 23\frac{11}{12} \\ -15\frac{9}{12} \\ \hline \end{array}$

Subtracting Fractions from Whole Numbers

To subtract a fraction from a whole number, *regroup* 1 from the whole number. Rewrite the 1 as a fraction with the same denominator as the fraction you're subtracting. Then subtract as usual.

Example 1. Subtract $\frac{5}{8}$ from 1.

Step 1. Rewrite 1 as $\frac{8}{8}$.

$$\begin{array}{cc} 1 & \frac{8}{8} \\ -\frac{5}{8} & -\frac{5}{8} \\ \hline & \frac{3}{8} \end{array}$$

Step 2. Subtract.

Answer: $\frac{3}{8}$

Example 2. Subtract $\frac{2}{3}$ from 5.

Step 1. Regroup 1 from 5. Write the 1 as $\frac{3}{3}$ and rewrite 5 as $4\frac{3}{3}$.

$$\begin{array}{cc} 5 & 4\frac{3}{3} \\ -\frac{2}{3} & -\frac{2}{3} \\ \hline & 4\frac{1}{3} \end{array}$$

Step 2. Subtract.

Answer: $4\frac{1}{3}$

Practice

Rewrite each whole number as indicated.

1. $1 = \frac{}{3}$ $\qquad$ $1 = \frac{}{4}$ $\qquad$ $1 = \frac{}{12}$ $\qquad$ $1 = \frac{}{2}$ $\qquad$ $1 = \frac{}{6}$

2. $3 = 2\frac{}{8}$ $\qquad$ $2 = 1\frac{}{3}$ $\qquad$ $5 = 4\frac{}{6}$ $\qquad$ $7 = 6\frac{}{2}$ $\qquad$ $12 = 11\frac{}{4}$

Subtract.

3.
$$\begin{array}{r} 1 \\ -\frac{3}{4} \\ \hline \end{array} \qquad \begin{array}{r} 1 \\ -\frac{5}{8} \\ \hline \end{array} \qquad \begin{array}{r} 1 \\ -\frac{7}{16} \\ \hline \end{array} \qquad \begin{array}{r} 1 \\ -\frac{1}{3} \\ \hline \end{array} \qquad \begin{array}{r} 1 \\ -\frac{1}{2} \\ \hline \end{array}$$

4.
$$\begin{array}{r} 2 \\ -\frac{1}{3} \\ \hline \end{array} \qquad \begin{array}{r} 5 \\ -\frac{3}{4} \\ \hline \end{array} \qquad \begin{array}{r} 3 \\ -\frac{5}{8} \\ \hline \end{array} \qquad \begin{array}{r} 7 \\ -\frac{3}{10} \\ \hline \end{array} \qquad \begin{array}{r} 6 \\ -\frac{5}{16} \\ \hline \end{array}$$

5.
$$\begin{array}{r} 4 \\ -\frac{3}{5} \\ \hline \end{array} \qquad \begin{array}{r} 9 \\ -\frac{2}{3} \\ \hline \end{array} \qquad \begin{array}{r} 8 \\ -\frac{7}{8} \\ \hline \end{array} \qquad \begin{array}{r} 11 \\ -\frac{7}{10} \\ \hline \end{array} \qquad \begin{array}{r} 12 \\ -\frac{11}{16} \\ \hline \end{array}$$

Subtracting Fractions by Regrouping

To subtract when the bottom fraction is larger than the top fraction, or when there is no top fraction, you regroup the top whole number.

Subtract	Step 1	Step 2	Step 3
$6\frac{1}{4}$	$5\frac{4}{4} + \frac{1}{4}$	$5\frac{5}{4}$	$5\frac{5}{4}$
$-2\frac{3}{4}$	$-2\frac{3}{4}$	$-2\frac{3}{4}$	$-2\frac{3}{4}$
			$3\frac{2}{4} = 3\frac{1}{2}$

Step 1. Regroup 6 as $5\frac{4}{4}$. So, $6\frac{1}{4} = 5\frac{4}{4} + \frac{1}{4}$.

Step 2. Combine the top fractions: $\frac{4}{4} + \frac{1}{4} = \frac{5}{4}$. So, $6\frac{1}{4} = 5\frac{5}{4}$.

Step 3. Subtract each column and reduce if needed: $3\frac{2}{4} = 3\frac{1}{2}$.

Answer: $3\frac{1}{2}$

Practice

■ **Subtract.**

1.
$$3\frac{1}{4} \qquad\qquad 2\frac{1}{3} \qquad\qquad 1\frac{3}{5}$$
$$-\ \frac{3}{4} \qquad\qquad -\ \frac{2}{3} \qquad\qquad -\ \frac{4}{5}$$

2.
$$5 \qquad\qquad 3 \qquad\qquad 5$$
$$-\ 2\frac{1}{2} \qquad\qquad -\ 1\frac{2}{3} \qquad\qquad -\ 3\frac{5}{8}$$

3.
$$4\frac{1}{6} \qquad\qquad 8\frac{5}{16} \qquad\qquad 7\frac{3}{8}$$
$$-\ 3\frac{5}{6} \qquad\qquad -\ 4\frac{11}{16} \qquad\qquad -\ 2\frac{7}{8}$$

4.
$$14\frac{1}{4} \qquad\qquad 19\frac{3}{8} \qquad\qquad 26\frac{5}{12}$$
$$-\ 8\frac{2}{4} \qquad\qquad -\ 10\frac{4}{8} \qquad\qquad -\ 18\frac{8}{12}$$

Adding and Subtracting Unlike Fractions

To add or subtract unlike fractions, rewrite them as like fractions. Then add or subtract as usual. (To review changing unlike fractions to like fractions, reread pages 56 and 57 at this time.)

Example 1. Add $\frac{1}{2}$ and $\frac{2}{3}$.

Step 1. Comparing multiples, choose 6 as the lowest common denominator; then write *like* fractions.

$$2: \; 2 \quad 4 \quad 6$$
$$3: \; 3 \quad 6 \quad 9$$

Step 2. Add. Simplify the answer if needed.

Step 1

$$\frac{1}{2} = \frac{3}{6}$$
$$+\frac{2}{3} = +\frac{4}{6}$$

Step 2

$$\frac{3}{6}$$
$$+\frac{4}{6}$$
$$\overline{\frac{7}{6}} = 1\frac{1}{6}$$

Answer: $1\frac{1}{6}$

Example 2. Subtract $2\frac{1}{2}$ from $7\frac{3}{8}$.

Step 1. Write *like* fractions with 8 as the lowest common denominator. (Because 8 is evenly divisible by 2, you do not need to compare multiples.)

Step 2. Since $\frac{4}{8} > \frac{3}{8}$, regroup 7 as $6\frac{8}{8}$.
Write $6\frac{8}{8} + \frac{3}{8}$ as $6\frac{11}{8}$; then subtract.

Step 1

$$7\frac{3}{8} = 7\frac{3}{8}$$
$$-2\frac{1}{2} = -2\frac{4}{8}$$

Step 2

$$7\frac{3}{8} = 6\frac{8}{8} + \frac{3}{8} = 6\frac{11}{8}$$
$$-2\frac{4}{8} \qquad\qquad = -2\frac{4}{8}$$
$$\overline{\qquad\qquad\qquad 4\frac{7}{8}}$$

Answer: $4\frac{7}{8}$

Practice

■ Add or subtract as indicated.

1.
$$\frac{1}{2} \qquad \frac{2}{3} \qquad \frac{3}{4} \qquad \frac{7}{8} \qquad \frac{7}{8}$$
$$+\frac{1}{4} \qquad +\frac{1}{6} \qquad +\frac{5}{8} \qquad +\frac{1}{2} \qquad +\frac{11}{16}$$

2.
$$\frac{9}{10} \qquad \frac{3}{4} \qquad \frac{5}{6} \qquad \frac{7}{8} \qquad \frac{15}{16}$$
$$-\frac{3}{5} \qquad -\frac{5}{8} \qquad -\frac{1}{3} \qquad -\frac{3}{4} \qquad -\frac{7}{8}$$

3.
$$\frac{2}{3} \qquad \frac{2}{3} \qquad \frac{3}{5} \qquad \frac{5}{6} \qquad \frac{1}{2}$$
$$+\frac{1}{4} \qquad -\frac{3}{5} \qquad +\frac{2}{3} \qquad -\frac{3}{4} \qquad -\frac{3}{7}$$

4.
$$\frac{3}{4} \qquad \frac{1}{2} \qquad \frac{4}{5} \qquad \frac{4}{5} \qquad \frac{5}{6}$$
$$-\frac{2}{3} \qquad +\frac{1}{3} \qquad -\frac{1}{2} \qquad +\frac{3}{4} \qquad +\frac{3}{4}$$

Adding Mixed Numbers

■ **As your first step, rewrite unlike fractions as like fractions. Add fractions and whole numbers separately; then simplify.**

5. $3\frac{1}{3}$ $5\frac{1}{4}$ $6\frac{3}{8}$ $4\frac{5}{12}$

     $+ 2\frac{1}{6}$ $+ 1\frac{1}{2}$ $+ 4\frac{1}{4}$ $+ 2\frac{1}{3}$

6. $5\frac{2}{3}$ $7\frac{3}{4}$ $3\frac{4}{5}$ $8\frac{11}{16}$

     $+ 2\frac{5}{6}$ $+ 5\frac{1}{2}$ $+ 3\frac{7}{10}$ $+ 7\frac{7}{8}$

7. $6\frac{3}{4}$ $8\frac{2}{3}$ $9\frac{3}{5}$ $14\frac{3}{5}$

     $+ 5\frac{2}{3}$ $+ 4\frac{1}{2}$ $+ 6\frac{2}{3}$ $+ 11\frac{3}{4}$

Subtracting Mixed Numbers

■ **As your first step, rewrite unlike fractions as like fractions. Regroup if needed; then subtract.**

8. $9\frac{7}{8}$ $5\frac{2}{3}$ $4\frac{11}{12}$ $13\frac{3}{4}$

     $- 5\frac{3}{4}$ $- 2\frac{1}{6}$ $- 1\frac{1}{4}$ $- 9\frac{1}{2}$

9. $8\frac{3}{8}$ $3\frac{1}{3}$ $7\frac{5}{12}$ $15\frac{1}{4}$

     $- 5\frac{3}{4}$ $- 1\frac{5}{6}$ $- 2\frac{3}{4}$ $- 10\frac{1}{2}$

10. $9\frac{1}{4}$ $6\frac{1}{3}$ $11\frac{1}{5}$ $23\frac{1}{3}$

     $- 4\frac{2}{3}$ $- 5\frac{1}{2}$ $- 3\frac{1}{2}$ $- 14\frac{4}{5}$

Applying Your Skills

Solve the following fraction addition and subtraction word problems.

1. Allie ran $2\frac{3}{4}$ miles on Monday, $3\frac{1}{4}$ miles on Wednesday, and $4\frac{3}{4}$ miles on Friday. How many total miles did Allie run this week?

2. Two wooden beams are placed side by side. One is $12\frac{15}{16}$ inches wide, and the other is $8\frac{7}{8}$ inches wide. What is the combined width of the two beams?

3. The clerk at Valley Fabric cut $4\frac{1}{3}$ yards of cloth off a bolt of fabric containing $9\frac{2}{3}$ yards. After she removed this piece, how much fabric was left on the bolt?

4. Connie ordered carpet that is $\frac{3}{4}$ inch thick. The accompanying pad is $\frac{5}{8}$ inch thick. How thick are the carpet and pad when measured together?

5. Joe's pickup is able to carry $\frac{7}{8}$ ton of rocks. Frieda's pickup can carry $\frac{3}{4}$ ton of rocks. How much more rock can Joe's pickup haul in each load than Frieda's?

6. At Twin Pines Lumber, Ed bought $2\frac{1}{4}$ pounds of #6 nails and $1\frac{2}{3}$ pounds of #8 nails. What total weight of nails did Ed buy?

7. Katrina had planned to buy $8\frac{1}{2}$ pounds of apples for the picnic. Instead, she bought a bargain bag that weighed $6\frac{2}{3}$ pounds. How many fewer pounds of apples did Katrina buy than she had planned?

8. During the first three weeks in November, Eugene recorded the following amounts of rainfall: week 1, $2\frac{3}{4}$ inches; week 2, $1\frac{2}{3}$ inches; week 3, $2\frac{7}{12}$ inches. How much rain fell during these three weeks?

9. A biscuit recipe calls for $2\frac{1}{4}$ cups of flour. If Lauren has only $\frac{7}{8}$ cup of flour, how much more does she need?

10. Meg's new curtain rod extends to a width of $51\frac{5}{8}$ inches. How much wider is the rod than Meg's $46\frac{13}{16}$-inch-wide bedroom window?

11. Following a recipe, Amanda mixed $2\frac{1}{2}$ cups of flour together with $\frac{3}{4}$ cup of sugar and $\frac{1}{3}$ cup of butter. She combined these ingredients in a bowl that holds 8 cups. How many more cups of ingredients can this bowl hold before it is full?

12. For presents, Alex makes serving trays as shown below.

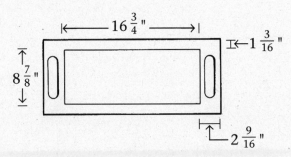

a) What are the total length and width of the tray Alex makes?

   length: _____ width: _____

b) How much longer is the tray than it is wide?

13. Find the total overtime hours worked on the 3-day weekend by each employee listed below.

Workday	J. Blakely	R. Craig
Saturday	$1\frac{1}{3}$ hr.	$\frac{3}{4}$ hr.
Sunday	$\frac{1}{2}$ hr.	$1\frac{1}{2}$ hr.
Monday	$2\frac{5}{6}$ hr.	$\frac{2}{3}$ hr.
Total:		

14. The newspaper said that 400 of the town's high school students had already had chicken pox. Of this group, $\frac{3}{5}$ contracted the virus before going into the 4th grade; $\frac{1}{3}$ got it while in the 4th, 5th, or 6th grade. What fraction of this group of students got chicken pox after the 6th grade?

15. As shown on the map below, Corner Grocery is between Will's house and Lisa's house.

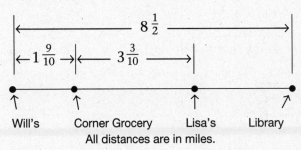

All distances are in miles.

a) How far does Lisa live from Will?

b) How far is the library from Corner Grocery?

16. On the first 3 weeks of her diet, Wanda lost weight as follows: week 1, $2\frac{1}{2}$ lb.; week 2, $1\frac{1}{4}$ lb.; and week 3, $\frac{7}{8}$ lb. Which equation below tells Wanda's average weekly weight loss?

(1) $y = (\frac{5}{2} + \frac{5}{4} + \frac{7}{8}) \times 3$

(2) $y = (\frac{5}{2} + \frac{5}{4} + \frac{7}{8}) \div \frac{1}{3}$

(3) $y = (\frac{5}{2} + \frac{5}{4} + \frac{7}{8}) \div 2$

(4) $y = (\frac{5}{2} + \frac{5}{4} + \frac{7}{8}) \div 3$

(5) None of the above

Problem Solver: Identifying Missing Information

On many math tests, some problems do not give all the necessary information you need to find a solution. Here are two ways these problems are written.

Not Enough Information Is Given

A. City Center Hardware advertised, "All nails, $1.78 per pound." Frank bought $7\frac{3}{4}$ pounds of #16 nails, $4\frac{1}{2}$ pounds of #12 nails, and a full box of #8 nails. How many pounds of nails did Frank purchase?

(1) $11\frac{1}{4}$ (4) $12\frac{4}{6}$

(2) $11\frac{4}{6}$ (5) Not enough information is given.

(3) $12\frac{1}{4}$

The correct answer choice is (5). To solve this problem, you must know the weight of a full box of #8 nails. The problem does not tell you this weight.

Before choosing "Not enough information is given," be careful! Be sure you can identify what information is missing. Only by doing this can you be sure this choice is correct.

What More Do You Need to Know?

B. City Center Hardware advertised, "All nails, $1.78 per pound." Frank bought $7\frac{3}{4}$ pounds of #16 nails, $4\frac{1}{2}$ pounds of #12 nails, and a full box of #8 nails. What more do you need to know to find out how many pounds of nails Frank purchased?

(1) the number of #8 nails Frank bought
(2) the regular price of #8 nails
(3) the weight of a full box of #8 nails
(4) the weight of a full box of #16 nails
(5) the weight of a full box of #12 nails

The correct answer choice is (3). As in problem A above, you need to know the weight of a full box of #8 nails before you can determine how many pounds of nails Frank purchased.

 Comparing problems A and B:

Similarities: Both give the same information. Both are concerned with finding how many pounds of nails Frank purchased.

Differences: Problem A checks your ability to recognize that the given information is incomplete. Problem B, on the other hand, asks you to identify missing information from a list of choices.

Practice

▨ **In each problem below, identify what information you need to know to answer the question.**

1. As part of her diet, Ellen counts the calories of foods she eats. How many calories will Ellen consume in a lunch consisting of a 485-calorie hamburger, a 175-calorie soft drink, and a serving of french fries?

   Missing information:

2. Twenty percent of the students in Mrs. Jones's preschool class are 3 years old, thirty-five percent are 4 years old, and the rest are either 5 or 6 years old. What total percent of Mrs. Jones's class is made up of 4- *and* 5-year-olds?

   Missing information:

3. One-fourth of Dave's salary goes for rent, one-third for food, and one-tenth for his car payment. What total fraction of his salary does Dave spend for rent, food, car payment, and utilities?

   Missing information:

4. When the price of sirloin steak was reduced from $4.49 per pound, Shauna bought 4 pounds. She also bought 3 pounds of hamburger for $1.39 per pound. What total amount did Shauna pay for these two items?

   Missing information:

5. To help pay for his new pickup, Lewis borrowed $8,000 from his credit union. He must pay back the loan in monthly payments of $208.75. To repay the loan, how much will Lewis actually pay the credit union?

   Missing information:

6. George bought $6\frac{3}{4}$ pounds of apples, a large bag of pears, and $7\frac{1}{2}$ pounds of peaches. How many more pounds of peaches did George buy than pears?

   Missing information:

7. Each Monday, Wednesday, and Friday evening, Lucas attends an adult math class. On these days, he studies for 2 hours each day. Each Sunday, Tuesday, and Thursday, Lucas spends additional time studying to prepare for the next day's class. How many hours does Lucas study math each week?

   Missing information:

Multiplying Fractions

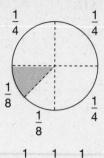

$$\tfrac{1}{2} \text{ of } \tfrac{1}{4} \text{ is } \tfrac{1}{8}$$

To multiply by a fraction is to find a part of something.

As shown at right, $\frac{1}{2}$ of $\frac{1}{4}$ is $\frac{1}{8}$.

Multiplying fractions is the same for both like fractions and unlike fractions.
- Multiply the numerators to find the numerator of the answer.
- Multiply the denominators to find the denominator of the answer.

Example 1. $\frac{3}{4} \times \frac{1}{2}$

Step 1. Multiply the numerators.

$$\frac{3}{4} \times \frac{1}{2} = \frac{3}{}$$

Step 2. Multiply the denominators.

$$\frac{3}{4} \times \frac{1}{2} = \frac{3}{8}$$

Answer: $\frac{3}{8}$

Example 2. $\frac{2}{3} \times \frac{3}{8}$

Step 1. Multiply the numerators.

$$\frac{2}{3} \times \frac{3}{8} = \frac{6}{}$$

Step 2. Multiply the denominators.

$$\frac{2}{3} \times \frac{3}{8} = \frac{6}{24}$$

Step 3. Reduce the answer.

$$\frac{6}{24} = \frac{6 \div 6}{24 \div 6} = \frac{1}{4}$$

Answer: $\frac{1}{4}$

Practice

Multiply. Reduce your answers.

1. $\frac{1}{2} \times \frac{2}{3} =$ $\frac{3}{4} \times \frac{1}{3} =$ $\frac{2}{3} \times \frac{7}{10} =$

2. $\frac{3}{5} \times \frac{1}{9} =$ $\frac{3}{8} \times \frac{1}{2} =$ $\frac{2}{5} \times \frac{7}{8} =$

3. $\frac{3}{4} \times \frac{2}{3} =$ $\frac{3}{7} \times \frac{3}{10} =$ $\frac{3}{4} \times \frac{5}{6} =$

4. $\frac{3}{4} \times \frac{3}{4} =$ $\frac{2}{4} \times \frac{3}{10} =$ $\frac{11}{12} \times \frac{3}{4} =$

5. $\frac{1}{3} \times \frac{3}{4} \times \frac{2}{5} =$ $\frac{2}{3} \times \frac{1}{2} \times \frac{3}{4} =$ $\frac{3}{5} \times \frac{2}{3} \times \frac{1}{4} =$

Using Canceling to Simplify Multiplication

When multiplying fractions, you can often use a shortcut called **canceling**.

To cancel, divide both a numerator and a denominator by the same number. The numerator and denominator can be in the same fraction or parts of two different fractions.

Canceling is similar to reducing a fraction. However, when you cancel, you simplify fractions *before* you multiply.

Example 1. $\frac{5}{6} \times \frac{3}{4}$

Step 1. Ask, "Can a numerator and a denominator be divided by the same number?" Yes. 6 and 3 can be divided by 3.

$$\frac{5}{6} \times \frac{3}{4}$$

Step 2. Divide: $6 \div 3 = 2$ and $3 \div 3 = 1$. Multiply the new fractions.

$$\frac{5}{\overset{}{6}} \times \frac{\overset{1}{3}}{4} = \frac{5}{2} \times \frac{1}{4} = \frac{5}{8}$$

Answer: $\frac{5}{8}$

Example 2. $\frac{4}{9} \times \frac{3}{8}$

Step 1. Divide both the 4 and the 8 by 4.

Step 2. Divide both the 9 and the 3 by 3.

$$\frac{\overset{1}{\cancel{4}}}{\underset{3}{\cancel{9}}} \times \frac{\overset{1}{\cancel{3}}}{\underset{2}{\cancel{8}}} = \frac{1}{3} \times \frac{1}{2} = \frac{1}{6}$$

Answer: $\frac{1}{6}$

Practice

▨ **Multiply. Use canceling as your first step.**

6. $\dfrac{5}{6} \times \dfrac{3}{4} =$ $\dfrac{2}{5} \times \dfrac{11}{12} =$ $\dfrac{7}{8} \times \dfrac{6}{7} =$

7. $\dfrac{3}{4} \times \dfrac{8}{9} =$ $\dfrac{1}{4} \times \dfrac{2}{5} =$ $\dfrac{5}{9} \times \dfrac{3}{10} =$

8. $\dfrac{5}{12} \times \dfrac{7}{15} =$ $\dfrac{3}{4} \times \dfrac{4}{15} =$ $\dfrac{7}{8} \times \dfrac{5}{7} =$

9. $\dfrac{3}{8} \times \dfrac{4}{6} =$ $\dfrac{1}{2} \times \dfrac{3}{4} \times \dfrac{8}{9} =$ $\dfrac{3}{4} \times \dfrac{4}{5} \times \dfrac{10}{12} =$

Multiplying Fractions, Whole Numbers, and Mixed Numbers

When you are multiplying any combination of fractions, whole numbers, and mixed numbers, the first step is to rename any whole number or mixed number as an improper fraction.

Example 1. $\frac{2}{3} \times 5$

Step 1. Rename the whole number 5 as a fraction. Write 5 as $\frac{5}{1}$.

Step 2. Multiply the fractions.

$$\frac{2}{3} \times \frac{5}{1} = \frac{10}{3} \quad \leftarrow \frac{\text{numerator times numerator}}{\text{denominator times denominator}}$$

Answer: $\frac{10}{3} = 3\frac{1}{3}$

Example 2. $3\frac{1}{2} \times \frac{1}{4}$

Step 1. Change $3\frac{1}{2}$ to an improper fraction.
  a) Change 3 to an improper fraction with a denominator of 2.

$$3 = \frac{6}{2}$$

  b) Add $\frac{6}{2}$ and $\frac{1}{2}$.

$$3\frac{1}{2} = \frac{6}{2} + \frac{1}{2} = \frac{7}{2}$$

Step 2. Multiply $\frac{7}{2}$ by $\frac{1}{4}$.

$$\frac{7}{2} \times \frac{1}{4} = \frac{7}{8}$$

Answer: $\frac{7}{8}$

Practice

▮ **Write each whole number as an improper fraction.**

1. $2 =$ $\qquad$ $4 =$ $\qquad$ $7 =$ $\qquad$ $5 =$ $\qquad$ $3 =$ $\qquad$ $12 =$

▮ **Write each mixed number as an improper fraction.**

2. $1\frac{2}{3} = \frac{3}{3} + \frac{2}{3} = \frac{5}{3}$ $\quad$ $3\frac{1}{4} =$ $\qquad$ $5\frac{3}{8} =$ $\qquad$ $4\frac{3}{5} =$ $\qquad$ $8\frac{1}{3} =$ $\qquad$ $11\frac{1}{2} =$

▮ **Multiply, using cancellation when possible. Simplify answers.**

3. $\frac{1}{2} \times 4 = \frac{1}{\cancel{2}} \times \frac{\cancel{4}^2}{1} = \frac{2}{1} = 2$ $\qquad$ $\frac{2}{3} \times 5 =$ $\qquad$ $6 \times \frac{2}{5} =$

4. $\frac{3}{4} \times 8 =$ $\qquad$ $\frac{1}{3} \times 9 =$ $\qquad$ $10 \times \frac{4}{5} =$

5. $12 \times \frac{2}{3} =$ $\qquad$ $20 \times \frac{4}{5} =$ $\qquad$ $\frac{5}{6} \times 18 =$

■ **Math Tip**

Some students like to use this shortcut for changing a mixed number to an improper fraction:

Numerator: Multiply the whole number by the denominator and add the numerator.

$$2\frac{3}{4} = \frac{(4 \times 2) + 3}{4} = \frac{11}{4} \qquad 5\frac{2}{3} = \frac{(3 \times 5) + 2}{3} = \frac{17}{3}$$

Denominator: Remains the same.

■ **Use the shortcut to change each mixed number to an improper fraction.**

6. $5\frac{1}{2} =$ $\qquad\qquad\qquad$ $2\frac{7}{8} =$ $\qquad\qquad\qquad$ $4\frac{3}{5} =$

■ **Multiply, using cancellation when possible. Simplify answers.**

7. $2\frac{2}{3} \times 4 =$ $\qquad\qquad$ $1\frac{3}{4} \times 3 =$ $\qquad\qquad$ $4\frac{1}{2} \times 6 =$

8. $6 \times 3\frac{1}{4} =$ $\qquad\qquad$ $4\frac{5}{12} \times 3 =$ $\qquad\qquad$ $6\frac{2}{5} \times 10 =$

9. $2\frac{1}{2} \times \frac{2}{3} =$ $\qquad\qquad$ $1\frac{5}{8} \times \frac{4}{5} =$ $\qquad\qquad$ $\frac{3}{4} \times 3\frac{1}{8} =$

10. $\frac{5}{8} \times 3\frac{1}{4} =$ $\qquad\qquad$ $4\frac{1}{2} \times \frac{5}{6} =$ $\qquad\qquad$ $\frac{4}{5} \times 1\frac{11}{12} =$

11. $3\frac{1}{2} \times 2\frac{2}{3} =$ $\qquad\qquad$ $4\frac{1}{4} \times 1\frac{3}{4} =$ $\qquad\qquad$ $5\frac{1}{2} \times 3\frac{5}{8} =$

12. $4\frac{1}{3} \times 3\frac{7}{8} =$ $\qquad\qquad$ $1\frac{15}{16} \times 2\frac{1}{4} =$ $\qquad\qquad$ $2\frac{3}{4} \times 5\frac{1}{4} =$

Dividing Fractions

Dividing is finding out how many times one amount can be found in a second amount. This is also true with fractions.

To find out how many $\frac{1}{4}$-inch-long sections are in a length of $1\frac{1}{2}$ inches, divide $1\frac{1}{2}$ by $\frac{1}{4}$.

As shown at right, there are six $\frac{1}{4}$s in $1\frac{1}{2}$.

Dividing fractions involves just one more step than multiplying fractions.

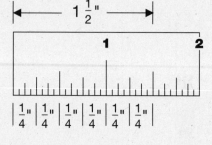

To divide fractions, *invert the divisor* (the number you're dividing by) and *change the division sign to a multiplication sign*. Then multiply.

$$1\frac{1}{2} \div \frac{1}{4} = 6$$

Inverting the Divisor

To **invert** means to turn a fraction upside down. When you invert a fraction, you switch top and bottom numbers.

Inverting $\frac{2}{3}$

$$\frac{2}{3} \diagdown \frac{3}{2}$$

Before inverting a divisor, you must change it to a fraction if it's not one already.
- Change a whole-number divisor to a fraction by placing the whole number over 1.
- Change a mixed-number divisor to an improper fraction.

Type of Divisor	Example	Writing the Divisor as a Fraction	Inverting the Divisor and Changing the Sign
proper fraction	$\frac{3}{4} \div \frac{7}{8} =$	$\frac{3}{4} \div \frac{7}{8} =$	$\frac{3}{4} \times \frac{8}{7} =$
improper fraction	$\frac{3}{5} \div \frac{3}{2} =$	$\frac{3}{5} \div \frac{3}{2} =$	$\frac{3}{5} \times \frac{2}{3} =$
whole number	$\frac{3}{8} \div 5 =$	$\frac{3}{8} \div \frac{5}{1} =$	$\frac{3}{8} \times \frac{1}{5} =$
mixed number	$\frac{15}{16} \div 2\frac{3}{8} =$	$\frac{15}{16} \div \frac{19}{8} =$	$\frac{15}{16} \times \frac{8}{19} =$

Practice

Invert each number below.

1. $\frac{2}{3}$ $\qquad$ $\frac{5}{8}$ $\qquad$ $\frac{5}{6}$ $\qquad$ $\frac{1}{2}$ $\qquad$ $\frac{3}{4}$

2. $\frac{4}{3} \rightarrow \frac{3}{4}$ $\qquad$ $\frac{7}{5}$ $\qquad$ $\frac{9}{4}$ $\qquad$ $\frac{5}{2}$ $\qquad$ $\frac{21}{16}$

3. 2 $\qquad$ 3 $\qquad$ 5 $\qquad$ 4 $\qquad$ 7

4. $2\frac{2}{3}$ $\qquad$ $3\frac{1}{4}$ $\qquad$ $4\frac{1}{3}$ $\qquad$ $1\frac{5}{8}$ $\qquad$ $2\frac{3}{16}$

Dividing Fractions by Fractions

Step 1. Invert the divisor—the fraction to the right of the division sign.

Step 2. Change the division sign to a multiplication sign; then multiply.

Invert the divisor.

Divide: $\frac{3}{4} \div \frac{2}{3}$

Change the sign.

$= \frac{3}{4} \times \frac{3}{2} = \frac{9}{8}$

Answer: $\frac{9}{8} = 1\frac{1}{8}$

Practice

▨ **Divide.**

5. $\frac{1}{2} \div \frac{3}{4} =$ $\qquad$ $\frac{3}{4} \div \frac{1}{3} =$ $\qquad$ $\frac{1}{3} \div \frac{3}{8} =$ $\qquad$ $\frac{2}{3} \div \frac{2}{4} =$

6. $\frac{7}{8} \div \frac{7}{8} =$ $\qquad$ $\frac{3}{6} \div \frac{1}{3} =$ $\qquad$ $\frac{11}{16} \div \frac{1}{4} =$ $\qquad$ $\frac{7}{8} \div \frac{1}{8} =$

7. $\frac{2}{3} \div \frac{1}{2} =$ $\qquad$ $\frac{7}{8} \div \frac{1}{2} =$ $\qquad$ $\frac{4}{5} \div \frac{4}{5} =$ $\qquad$ $\frac{5}{16} \div \frac{1}{16} =$

8. $\frac{12}{8} \div \frac{2}{3} =$ $\qquad$ $\frac{14}{6} \div \frac{1}{2} =$ $\qquad$ $\frac{3}{4} \div \frac{5}{4} =$ $\qquad$ $\frac{4}{3} \div \frac{3}{2} =$

Dividing Fractions, Whole Numbers, and Mixed Numbers

When you are dividing any combination of fractions, mixed numbers, and whole numbers, the first step is to rewrite any mixed number or whole number as an improper fraction.

Dividing with Fractions and Whole Numbers

When you are dividing fractions with whole numbers
- write the whole number as a fraction with a denominator of 1
- invert the divisor; then multiply the fractions

Example 1. $\frac{3}{4} \div 7$

Invert the divisor.

$\frac{3}{4} \div 7 = \frac{3}{4} \div \frac{7}{1}$

Change the sign.

$= \frac{3}{4} \times \frac{1}{7} = \frac{3}{28}$

Answer: $\frac{3}{28}$

Example 2. $4 \div \frac{2}{3}$

Invert the divisor.

$4 \div \frac{2}{3} = \frac{4}{1} \div \frac{2}{3}$

Change the sign.

$= \frac{4}{1} \times \frac{3}{2} = \frac{12}{2} = 6$

Answer: 6

Practice

Divide.

1. $\frac{5}{8} \div 2 =$ 　　　 $\frac{3}{4} \div 4 =$ 　　　 $\frac{2}{3} \div 5 =$ 　　　 $\frac{5}{8} \div 5 =$

2. $\frac{11}{2} \div 4 =$ 　　　 $\frac{15}{12} \div 2 =$ 　　　 $\frac{3}{2} \div 3 =$ 　　　 $\frac{4}{3} \div 6 =$

3. $6 \div \frac{3}{4} =$ 　　　 $12 \div \frac{4}{3} =$ 　　　 $4 \div \frac{2}{3} =$ 　　　 $7 \div \frac{3}{8} =$

4. $8 \div \frac{2}{3} =$ 　　　 $3 \div \frac{1}{4} =$ 　　　 $4 \div \frac{3}{8} =$ 　　　 $15 \div \frac{2}{3} =$

5. $7 \div \frac{14}{3} =$ 　　　 $\frac{12}{7} \div 4 =$ 　　　 $3 \div \frac{9}{6} =$ 　　　 $\frac{4}{3} \div 6 =$

Dividing with Mixed Numbers

When you are dividing with mixed numbers
- change each mixed number to an improper fraction
- invert the divisor; then multiply the fractions

Example 1. $3\frac{1}{4} \div \frac{1}{2}$

$$3\frac{1}{4} \div \frac{1}{2} = \frac{13}{4} \div \frac{1}{2}$$

Invert, Change

$$= \frac{13}{4} \times \frac{2}{1}$$
$$= \frac{13}{2}$$
$$= 6\frac{1}{2}$$

Example 2. $2\frac{3}{8} \div 4$

$$2\frac{3}{8} \div 4 = \frac{19}{8} \div \frac{4}{1}$$

Invert, Change

$$= \frac{19}{8} \times \frac{1}{4}$$
$$= \frac{19}{32}$$

Example 3. $1\frac{1}{3} \div 2\frac{1}{2}$

$$1\frac{1}{3} \div 2\frac{1}{2} = \frac{4}{3} \div \frac{5}{2}$$

Invert, Change

$$= \frac{4}{3} \times \frac{2}{5}$$
$$= \frac{8}{15}$$

Practice

Divide.

6. $1\frac{2}{5} \div \frac{1}{2} =$ $2\frac{3}{4} \div \frac{2}{3} =$ $\frac{3}{4} \div 1\frac{3}{8} =$ $\frac{3}{5} \div 3\frac{1}{2} =$

7. $3\frac{1}{2} \div 3 =$ $1\frac{3}{4} \div 4 =$ $2\frac{5}{8} \div 2 =$ $4\frac{2}{3} \div 5 =$

8. $2\frac{1}{2} \div 1\frac{1}{2} =$ $4\frac{1}{4} \div 8\frac{1}{2} =$ $2\frac{1}{4} \div 3\frac{3}{8} =$ $6\frac{3}{4} \div 2\frac{3}{8} =$

9. $\frac{7}{8} \div 2 =$ $2\frac{3}{4} \div 2 =$ $3 \div 4\frac{1}{2} =$ $4\frac{2}{3} \div 2\frac{1}{12} =$

Applying Your Skills

Solve the following multiplication and division word problems.

1. The paving crew has $\frac{7}{10}$ mile of road to pave by May 14th. The crew hopes to complete $\frac{2}{3}$ of this by May 6th. What length of road does the crew hope to finish by May 6?

2. Nora plans to use washers as spacers to raise the corner of an uneven table by $\frac{5}{8}$ inch. How many $\frac{1}{16}$-inch-thick washers will Nora need to level the table?

3. A cookbook recommends $\frac{1}{3}$ hour of cooking time per pound to cook a roast medium-well-done. Following these instructions, how long should a roast weighing $8\frac{1}{2}$ pounds be cooked if the chef wants it medium-well-done?

4. To make curtains for their home, the Brechts bought a $17\frac{1}{2}$-yard bolt of fabric. If it takes $3\frac{1}{4}$ yards of fabric for each window curtain, how many windows do the Brechts have enough fabric for? (Hint: Your answer must be a whole number!)

5. Trish, a housepainter, painted $6\frac{1}{2}$ wall panels in one hour. At this rate, how many *complete* wall panels can she expect to paint in $7\frac{3}{4}$ hours?

6. For the barbecue, Mike bought $4\frac{1}{2}$ pounds of hamburger. If he divides this into patties weighing about $\frac{1}{4}$ pound, how many patties can he make?

7. Wallpaper comes in rolls that are $1\frac{3}{4}$ feet wide. How many side-by-side strips of wallpaper will Ray need to paper the wall of his den, which is 15 feet wide?

8. A carpenter cut a $3\frac{1}{4}$-foot piece off a 12-foot length of fir molding. The remaining piece was then cut into 7 equal lengths. Assuming no waste, how long is each of these 7 pieces?

9. Spencer built a picnic table, the end view of which is shown below. How wide is the table?
(Hint: Width = 6 table boards + 5 gaps.)

End View

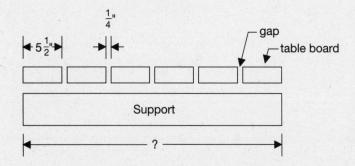

10. Ching-Yan bought $1\frac{1}{4}$ pounds of cheese on sale for $3 per pound and a $6\frac{1}{2}$-pound package of sirloin steaks on sale for $4 per pound. Which expression at right shows how much Ching-Yan will pay in all for these items?

(1) $(\frac{5}{4} + \$3) + (\frac{13}{2} + \$4)$

(2) $(\frac{5}{4} + \frac{13}{2}) + (\$3 + \$4)$

(3) $(\frac{5}{4} + \frac{13}{2})(\$3 + \$4)$

(4) $(\frac{5}{4})(\$4) + (\frac{13}{2})(\$3)$

(5) $(\frac{5}{4})(\$3) + (\frac{13}{2})(\$4)$

■ **Problems 11–12 refer to the circle graph at right.**

11. What fraction of their monthly income do the Nelsons spend on each of the following? Reduce each answer to lowest terms.

Rent: _____ Food: _____ Auto: _____

Nelson Family Budget
Monthly Income: $1,600

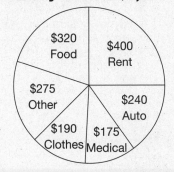

12. If the Nelsons donate $\frac{1}{25}$ of their income to charity, how much do they donate each month?

13. On Tuesday, stock values rose when the Federal Reserve announced a lowering of interest rates. Three stocks whose values changed are shown at right.

Note: Each increase stands for the fraction of a dollar that a stock's value increased.

What is the *average* increase in value of the three stocks listed?

Stock Symbol	Change in Value
ARI	$+\frac{3}{4}$
UMB	$+\frac{1}{2}$
RWM	$+\frac{7}{8}$

14. Jackie, a jeweler, uses various silver alloys to make earrings, bracelets, and buckles. Fill in the blank lines on the table to show how many complete items of each type she can make with her remaining alloy supply.

Item	Alloy Type	Alloy Supply in Ounces (oz.) (a)	Alloy Needed per Item (oz.) (b)	Number of Items That Can Be Made (a ÷ b)
Earrings	90/10	7	$\frac{3}{4}$	_____
Bracelets	80/20	$15\frac{1}{2}$	3	_____
Buckles	70/30	$22\frac{1}{2}$	$3\frac{2}{3}$	_____

Relating Fractions and Ratios

A **ratio** is a comparison of two numbers. For example, if there are 8 women and 5 men in your math class, the ratio of women to men is "8 to 5."

You can write the ratio "8 to 5" using symbols in two ways:
- As a fraction, write 8 to 5 as $\frac{8}{5}$.
- With a colon, write 8 to 5 as 8:5.

In words, always read a ratio with the word *to*.
Read the ratios $\frac{8}{5}$ and 8:5 as "eight to five."

To simplify a ratio, write it as a fraction and follow these four rules.

1. Reduce a ratio to lowest terms. 6 to 10 $= \frac{6}{10} = \frac{3}{5}$ **Answer:** $\frac{3}{5}$	3. Do not write an improper-fraction ratio as a mixed number. 12 to 8 $= \frac{12}{8} = \frac{3}{2}$ **Answer:** $\frac{3}{2}$ (Leave as an improper fraction.)
2. Write a whole-number ratio with a denominator of 1. 4 to 1 $= \frac{4}{1}$ (Leave as is.) **Answer:** $\frac{4}{1}$ 9 to 3 $= \frac{9}{3} = \frac{3}{1}$ **Answer:** $\frac{3}{1}$	4. Rewrite a complex ratio as an equal fraction that has only whole numbers. $5\frac{1}{2}$ to 3 $= \frac{5\frac{1}{2}}{3}$ To rewrite this ratio, divide the top mixed number ($5\frac{1}{2}$) by the bottom number (3). $5\frac{1}{2} \div 3 = \frac{11}{2} \div \frac{3}{1} = \frac{11}{2} \times \frac{1}{3} = \frac{11}{6}$ **Answer:** $\frac{11}{6}$

Practice

■ Write each ratio in lowest terms.

1. 4 to 6   6 to 8   15 to 3   6 to 4   8 to 16

2. 6 to 2   12 to 8 $4\frac{1}{2}$ to 3 $3\frac{2}{3}$ to 2 $6\frac{1}{2}$ to $4\frac{1}{3}$

To solve a ratio word problem, write the ratio in the same order as it appears in the question.	
One-Step Ratio Problem	**Two-Step Ratio Problem**
Wanda earns $1,200 per month. If she pays $300 each month in rent, what is the ratio of her *rent to* her *income?* Write the ratio of *rent to income.* Reduce. rent to income $= \dfrac{\text{rent}}{\text{income}} = \dfrac{\$300}{\$1,200}$ $= \dfrac{1}{4}$ Answer: $\dfrac{1}{4}$ *or* **1:4** *or* **1 to 4** Note: If the question had asked for the ratio of *income to rent,* the answer would be 4 to 1.	On a test of 40 questions, Thurman answered all of the questions and got 6 wrong. What is the ratio of his *correct* answers *to* his *incorrect* answers? *Step 1.* Determine how many questions Thurman answered correctly: $40 - 6 = 34$ correct. *Step 2.* Write the ratio of *correct to incorrect.* $\dfrac{\text{correct answers}}{\text{incorrect answers}} = \dfrac{34}{6} = \dfrac{17}{3}$ Answer: $\dfrac{17}{3}$ *or* **17:3** *or* **17 to 3**

■ **Solve the following ratio problems.**

3. A new compact car gets 48 miles to the gallon during highway driving and 32 miles to the gallon during city driving.
   a) What is the ratio of this car's city mileage to highway mileage?
   b) What is the ratio of this car's highway mileage to city mileage?

4. Last season the Bulldog basketball team won only 8 of its 22 games. It lost the rest.
   a) What is the ratio of the number of games the team won to the number of games it played?
   b) What is the ratio of the number of games the team won to the number of games it lost?

5. The Garrett Co. schedules 20 working days each month for each employee. On average, each employee misses $1\frac{1}{2}$ days each month due to illness. What is the average ratio of sick days to working days for employees at Garrett?

6. In an aerobics class with 30 students, there are 16 women and 14 men.
   a) What is the ratio of men to women in this class?
   b) What is the ratio of women to men?
   c) What is the ratio of the number of women to the total number of students?

7. One yard is equal to 3 feet (36 inches). To find a ratio of lengths, both lengths must be expressed in the same unit.
   a) What is the ratio of one yard to one foot? (Hint: As your first step, write 1 yard as a number of feet.)
   b) What is the ratio of 8 inches to 2 yards? (Hint: As your first step, write 2 yards as a number of inches.)

8. The road crew has paved $3\frac{1}{2}$ miles of the 8 miles of new road being put in at the airport.
   a) What is the ratio of the amount of paved road to the total amount of new road?
   b) What is the ratio of the amount of paved road to the amount of road still unpaved?

Life Skill: Working with Rates

Closely related to ratios are **rates**. Like a ratio, a rate is a comparison of two numbers. A rate, though, always compares one quantity to a unit (single amount) of another quantity.

Rates are usually expressed with the word *per*. Per means "for each."

You may already be familiar with several commonly used rates:

speed $= \dfrac{\text{miles}}{\text{hour}}$ (miles per hour)   **mileage** $= \dfrac{\text{miles}}{\text{gallon}}$ (miles per gallon)

unit price $= \dfrac{\text{dollars}}{\text{pound}}$ (dollars per pound)   **heart rate** $= \dfrac{\text{beats}}{\text{minute}}$ (beats per minute)

Often, you are asked to find a rate from a given ratio.

Example 1	Example 2
Aurora drove 171 miles in 3 hours. On the average, how many miles did Aurora drive each hour?	At Hong's Oriental Market, Li bought 4 pounds of duck for $11.56. What price was Li charged per pound?
The question is asking you to find Aurora's average rate of *speed*—the average number of miles she drove in 1 hour.	This question asks you to find the *unit price* that Li is paying—the amount she is paying for 1 pound of duck.
To find this rate, follow these steps:	To find this rate, follow these steps:

Step 1. Write a ratio comparing miles to hours.

$$\frac{\text{miles}}{\text{hour}} = \frac{171 \text{ miles}}{3 \text{ hours}}$$

Step 2. Find the rate by dividing the numerator (171) by the denominator (3).

$$\frac{171 \text{ miles}}{3 \text{ hours}} = \frac{57 \text{ miles}}{1 \text{ hour}}$$

Remember: A rate in fraction form has a denominator of 1. The rate $\frac{57 \text{ miles}}{1 \text{ hour}}$ can also be expressed as 57 miles per hour *or* 57 mph.

Answer: 57 miles per hour *or* 57 mph

Step 1. Write a ratio comparing price to pounds.

$$\frac{\text{price}}{\text{pounds}} = \frac{\$11.56}{4 \text{ pounds}}$$

Step 2. Find the unit price by dividing $11.56 by 4.

$$\frac{\$11.56}{4 \text{ pounds}} = \frac{\$2.89}{1 \text{ pound}}$$

Unit Price: $2.89 per pound *or* $2.89/lb.

Practice

▪ Find the rates in the following problems.

1. Over a 5-minute period, Ben counted his heartbeats. He counted 380 beats. At what rate (beats per minute) is Ben's heart beating?

2. A high-speed train in Europe traveled 570 miles in 3 hours. What was the average speed of this train during the trip?

3. Valley Car Rental has a "5-day special" car rental package for $165. What is the daily rate that Valley is charging on this package?

4. Rosie rode her bike from Bay City to Evans, a distance of 72 miles. The trip took $4\frac{1}{2}$ hours. What was Rosie's average speed for this distance?

5. The results of a gas mileage (miles per gallon) test performed on four cars is shown below. Compute the gas mileage of each car. Write your answers on the table.

Car	Miles Driven	Gas Used	Gas Mileage
#1	144	6 gal.	
#2	195	$7\frac{1}{2}$ gal.	
#3	65	$3\frac{1}{4}$ gal.	
#4	78	$4\frac{1}{3}$ gal.	

6. During a rainstorm, $1\frac{1}{4}$ inches of rain fell in Lakeview in a $4\frac{1}{2}$-hour period. What was the average rate of rainfall in inches per hour during this storm?

7. Marlena earned $40.25 for $3\frac{1}{2}$ hours of overtime work on Saturday. Determine Marlena's overtime pay rate.

8. Abdulla walked $8\frac{3}{4}$ miles in $2\frac{1}{2}$ hours. Expressing your answer in miles per hour, determine Abdulla's walking rate.

9. The Community Wading Pool was drained so that repairs could be made. It took 3 hours and 20 minutes to drain the pool's 15,000 gallons of water. What was the drainage rate in gallons per hour? (Hint: Write 20 minutes as a fraction of an hour.)

10. To the nearest cent, compute the price per ounce of each size of soft drink shown below.

a) _____ per oz.   c) _____ per oz.

$0.59   8 oz. $0.99   16 oz.

b) _____ per oz.   d) _____ per oz.

$0.74   12 oz. $1.29   24 oz.

Problem Solver: Recognizing and Comparing Patterns

Recognizing Patterns

You can solve many types of problems by recognizing a pattern. First, you determine the common link that relates numbers in the pattern. Then, you can use that link to continue a **number series** or to predict a future value based on a **trend** in data.

Number Series

What is the next number in the following number series?

$$6, 12, 24, \ldots$$

Step 1. Find the common link.

$$\times 2 \quad \times 2 \quad \times 2$$
$$6 \quad 12 \quad 24 \quad ?$$

Link: Multiply by 2.

Step 2. Find the next number.

Multiply 24 by 2: $24 \times 2 = 48$

Next number: 48

Data Trend

The value of Scott Computer Company stock has risen as shown below. End-of-the-month values are given. If this pattern (or trend) continues, what will the stock's value be at the end of July?

Month: April   May   June   July

Value: $55\frac{1}{4}$ $55\frac{1}{2}$ $55\frac{3}{4}$?

Step 1. Find the common link.

$$+\frac{1}{4} \quad +\frac{1}{4} \quad +\frac{1}{4}$$
$$55\frac{1}{4} \quad 55\frac{1}{2} \quad 55\frac{3}{4} \quad ?$$

Link: Add $\frac{1}{4}$.

Step 2. Find the predicted July value.

Add $\frac{1}{4}$ to $55\frac{3}{4}$: $55\frac{3}{4} + \frac{1}{4} = 56$

Predicted July value: 56

Comparing Patterns

Sometimes two or more patterns can be found in a problem. Comparing these patterns as rows on a table may help you answer questions about the given information.

> Laura plans to buy one of two used cars. The Honda costs $12,000 and will depreciate (lose value) at the rate of about $1,500 each year. The Ford costs $13,500 and will depreciate at the rate of about $2,000 per year. In how many years will the value of the cars be the same?

Write a table where the rows are the *depreciation patterns* described in the problem. Compare the rows year by year until the two values become equal.

	Value Now	After Year 1	After Year 2	After Year 3
Honda	$12,000	$10,500	$9,000	$7,500
Ford	$13,500	$11,500	$9,500	$7,500

Answer: The cars will have the same value ($7,500) in **three years**.

Practice

▨ Identify the patterns.

1. What is the next number in this series?
   $5, 8\frac{1}{2}, 12, 15\frac{1}{2}, \ldots$

3. Find the next term in the following series.
   $2, 6, 18, 54, \ldots$

2. Which series below is described by the
   following *number-to-number* rule?
   "times two and subtract 3"

   (1) 5, 7, 14, 20, . . .
   (2) 5, 7, 11, 19, . . .
   (3) 5, 8, 13, 17, . . .
   (4) 5, 8, 15, 19, . . .
   (5) 5, 7, 14, 23, . . .

4. Choose the *alternate-number* rule that best
   tells how the following series is changing.
   6, 12, 9, 15, 12, 18, 15, . . .

   (1) add 6, add 3
   (2) subtract 6, add 3
   (3) add 4, subtract 3
   (4) add 6, subtract 3
   (5) add 3, subtract 6

▨ Use the tables to compare patterns.

5. Pam has her choice of two jobs. The job at The Valley Restaurant (TVR) pays $15,000 a year with
   yearly raises of $750. The job at Norman's Office Supplies (NOS) pays $13,650 a year with yearly
   raises of $1,200. In how many years will the annual salaries be equal? (Write the *salary patterns* on
   the table shown below. Compare year by year.)

Starting Salary	After 1 Year	After 2 Years	After 3 Years	After 4 Years	After 5 Years
TVR: $15,000					
NOS: $13,650					

6. As shown on the table below, plant A and plant B had different heights when growth
   measurements were started. If the weekly growth rate of each plant stays constant, at the end of
   which week will the two plants be the same height?
   (Write the *growth patterns* on the table below. Compare week by week.)

Starting Height	After 1 Week	After 2 Weeks	After 3 Weeks	After 4 Weeks	After 5 Weeks
Plant A: $6\frac{3}{4}$ in.	$7\frac{1}{2}$ in.				
Plant B: $6\frac{1}{4}$ in.	$7\frac{1}{8}$ in.				

Data Highlight: **Reading a Pictograph**

A pictograph uses pictures, or symbols, to graph information. The value of a symbol is defined in a key on the graph.

   To find the total value of a line of symbols, multiply the number of symbols by the value of a single symbol. Half of a symbol has half the value of a complete symbol.

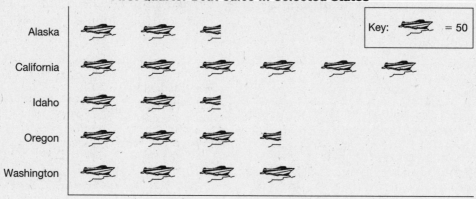

EMERALD VALLEY BOAT COMPANY
First Quarter Boat Sales in Selected States

Examples

A. For the sales period shown, how many boats did Emerald Valley sell in Oregon?

B. What is the *ratio* of sales in California to the sales in Washington?

C. What is the *average* number of boats sold per state for the five states listed?

Answers

A. *Step 1.* Count the number of boat symbols to the right of Oregon: $3\frac{1}{2}$.

   Step 2. Multiply $3\frac{1}{2}$ by 50, the value of each symbol.

   Answer: 175 boats

B. *Step 1.* Determine the number of boats sold in each state: California = 300, Washington = 200.

   Step 2. Write a ratio of California sales to Washington sales. Reduce this ratio.

   $$\text{Ratio} = \frac{300}{200} = \frac{3}{2}$$

C. *Step 1.* Count the total number of boat symbols shown for the five states: $18\frac{1}{2}$.

   Step 2. Divide: $18\frac{1}{2} \div 5 = 3\frac{7}{10}$ symbols per state

   Step 3. Multiply: $3\frac{7}{10} \times 50 = 185$ boats per state
   Answer: an average of 185 boats sold per state

Practice

■ **Problems 1–8 are based on the following pictograph.**

NEW ENGLAND BICYCLE COMPANY
Spring Season Sales Report

Key: = 150

Road Classic

Pro Sport

Touring Special

Mountain Ranger

All Terrain

1. Which bicycle model sold the most during the spring season?

2. About how many *Mountain Ranger* bicycles did the New England Bicycle Company sell during the spring season?

3. About how many more *Pro Sports* were sold this spring season than *All Terrains?*

4. If *Road Classics* sell for $328, about how much sales revenue was received during the spring season from the sale of this model?

5. Which two bicycle models sold about the same number during the spring season?

6. Approximately how many bicycles did the New England Bicycle Company sell in all during its spring season?

7. If the spring season lasts 3 months, what is the *average* number of sales per month of the *Pro Sport* model?

8. What is the *ratio* of the number of *Road Classics* sold to the number of *Touring Specials* sold?

■ **Solve the following problem.**

9. The *All Terrain* model sells for $295, and the *Pro Sport* model sells for $445. What more do you need to know to determine how much more profit is made on each *Pro Sport* sale than on each *All Terrain* sale?

   (1) the prices of each of the other three models of bicycles
   (2) the total yearly sales of *Pro Sport* and *All Terrain* models
   (3) the total yearly sales of all five bicycle models
   (4) the cost of making each *Pro Sport* and each *All Terrain* model
   (5) the weight of both the *Pro Sport* and *All Terrain* models

Test Readiness Checkup

■ **Circle each correct answer. Check your answers on page 244; then correct any errors.**

1. A shaft $8\frac{7}{8}$ inches long is laid end to end with a second shaft that is $14\frac{1}{4}$ inches long. Which of the following is the best estimate of the combined length of the two shafts?

   (1) 21 in.   (4) 24 in.
   (2) 22 in.   (5) 25 in.
   (3) 23 in.

■ **Problem 5 refers to the map below.**

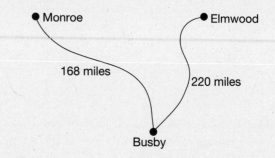

2. A $12\frac{7}{8}$-ton railroad car is loaded with 18 pickup trucks each weighing $2\frac{1}{4}$ tons. Which expression below gives the best estimate of the combined weight in tons of the railroad car and the pickups?

   (1) $18 \times (13 - 2)$ (4) $13 + 18 \times 2$
   (2) $12 + 18 \times 2$ (5) $18 \times 2 - 13$
   (3) $(13 + 18) \times 2$

5. After driving $\frac{1}{2}$ of the distance from Busby to Elmwood, how many miles is Kathleen from Elmwood?

   (1) 34   (4) 110
   (2) 66   (5) 126
   (3) 84

3. Stacey grew $1\frac{1}{4}$ inches two years ago, $1\frac{3}{8}$ inches last year, and $\frac{7}{8}$ inch so far this year. How many inches has Stacey grown during all this time?

   (1) $2\frac{11}{20}$ (4) 4

   (2) $3\frac{1}{2}$ (5) $4\frac{1}{4}$

   (3) $3\frac{7}{8}$

6. Denny mixes $\frac{3}{8}$ pint of thinner into each gallon of stain he uses. How many pints of thinner will Denny need for a house requiring 13 gallons of stain?

   (1) 1   (4) 8
   (2) 3   (5) 11
   (3) 5

4. The value of James Company stock fell from $16\frac{1}{2}$ to $14\frac{7}{8}$ between June 1 and June 15. How many points did the stock value drop during the two-week period?

   (1) $1\frac{5}{8}$ (4) 2

   (2) $1\frac{3}{4}$ (5) $2\frac{3}{8}$

   (3) $1\frac{7}{8}$

7. Bryan, a jewelry maker, uses $\frac{5}{16}$ ounce of gold for each ring he designs. How many complete rings can Bryan make when his gold supply is down to $2\frac{5}{8}$ ounces?

   (1) 6   (4) 9
   (2) 7   (5) 10
   (3) 8

8. Of every 10 people who order burgers at Chuck's Place, 5 order single burgers, 3 order double burgers, and 2 order cheeseburgers. What is the ratio of single burger sales to cheeseburger sales?

   (1) 1 to 5
   (2) 2 to 5
   (3) 2 to 1
   (4) 5 to 3
   (5) 5 to 2

9. During the first day of fall registration, 2,600 students enrolled at Walton College. Registration was completed in $6\frac{1}{2}$ hours. Expressed in *students per hour,* what was the registration rate during this first day?

   (1) 400 students per hour
   (2) 450 students per hour
   (3) 500 students per hour
   (4) 550 students per hour
   (5) Not enough information is given.

Problem 10 refers to the information below.

Birth Height	Week 1	Week 2
$17\frac{3}{4}$ in.	$18\frac{5}{8}$ in.	$19\frac{1}{2}$ in.

10. As shown above, Willie is keeping a record of his daughter's growth rate. If his daughter Amanda keeps growing at this same rate, what will be her height at the end of the *4th week?*

    (1) $20\frac{1}{8}$ in.
    (2) $20\frac{3}{8}$ in.
    (3) $20\frac{7}{8}$ in.
    (4) $21\frac{1}{4}$ in.
    (5) 22 in.

Problem 11 refers to the bus schedule below.

```
            BUS SCHEDULE
      Tuesday—Roosevelt Station
          Departure Times

            10:00 A.M.
            10:25 A.M.
            10:35 A.M.
            11:00 A.M.
            11:10 A.M.
            11:35 A.M.
```

11. Jonas found part of a bus schedule as shown above. If it is now Tuesday, 11:50 A.M., when can Jonas catch the next bus at this station? (Assume that the next several departure times follow the pattern shown.)

    (1) 11:45 A.M.   (4) 12:20 P.M.
    (2) 11:55 A.M.   (5) 12:35 P.M.
    (3) 12:10 P.M.

Problem 12 refers to the pictograph below.

CROSBY DOLL COMPANY
December Sales Report

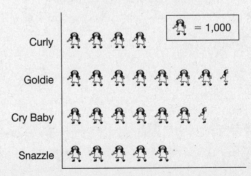

12. According to the pictograph, what is the ratio of the sales of *Goldie* dolls to the sales of *Snazzle* dolls?

    (1) 3 to 2   (4) 2 to 1
    (2) 4 to 3   (5) 3 to 1
    (3) 5 to 4

UNIT 5 Decimals

Estimating: Building Confidence with Decimals

Suppose a test question requires you to solve the following: 26.04 − (4.83 + 6.075 + 1.6). Your answer choices are shown at right. This looks like a tough question! But estimation makes it easy. Simply subtract the sum of 5 + 6 + 2 from 26, or 26 − 13 = 13. From the choices given, you quickly (and correctly) choose (3) 13.535 as the answer.

(1) 6.945
(2) 9.005
(3) 13.535
(4) 20.945
(5) 26.875

Estimation is useful anytime you don't need an exact answer. It can also help you choose an answer to a multiple-choice question as in the example above. And estimation can help when you are not sure where to place the decimal point in an exact answer.

Rounding Mixed Decimals

A **mixed decimal** is a whole number together with a decimal fraction. When you read a mixed decimal, read the decimal point as the word *and*. 5.83 is read as "five *and* eighty-three hundredths."

One way to estimate with mixed decimals is simply to round each mixed decimal to the nearest whole number. Then do the indicated math.

To round a mixed decimal to a whole number, look at the digit in the tenths place.
- If the digit is 5 or more, round up to the next larger whole number.
- If the digit is less than 5, drop all decimal digits. The whole number remains the same.

Math Tip
When rounding to a whole number, concentrate only on the digit in the tenths place.

┌─ tenths place
14.6 rounds to 15
└─ 5 or more
14.6 ≈ 15

┌─ tenths place
8.473 rounds to 8
└─ less than 5
8.473 ≈ 8

┌─ tenths place
6.54 rounds to 7
└─ 5 or more
6.54 ≈ 7

Example 1. Estimate: 7.53 + 5.9
Step 1. Round each mixed decimal to the nearest whole number.
　　7.53 ≈ 8　　5.9 ≈ 6
Step 2. Add the whole numbers.
　　8 + 6 = 14
Answer: 7.53 + 5.9 ≈ 14

Example 2. Estimate: 4.07 × 2.53
Step 1. Round each mixed decimal to the nearest whole number.
　　4.07 ≈ 4　　2.53 ≈ 3
Step 2. Multiply the whole numbers.
　　4 × 3 = 12
Answer: 4.07 × 2.53 ≈ 12

104

Practice

■ Round each mixed number to the nearest whole number or dollar.

1. a) 3.87 meters ≈   b) $9.83 ≈   c) 7.13 centimeters ≈

2. a) $24.17 ≈   b) 5.75 ounces ≈   c) 29.2 miles per gallon ≈

■ Estimate an answer for each problem below.

3.
$$\begin{array}{r} \text{Estimate} \\ \$7.81 \\ +\ 2.96 \\ \hline \end{array}$$
$$\begin{array}{r} \text{Estimate} \\ 8.4 \\ -\ 5.725 \\ \hline \end{array}$$
$$\begin{array}{r} \text{Estimate} \\ \$9.27 \\ \times\ 19.5 \\ \hline \end{array}$$

4.
$$\begin{array}{r} \text{Estimate} \\ 9.75 \\ \times\ 0.95 \\ \hline \end{array}$$
$$2.2\,\overline{)16.434} \quad \text{Estimate}$$
$$3.14\,\overline{)42.076} \quad \text{Estimate}$$

■ Use an estimate to help you decide where the decimal point belongs in each answer below.

5. 8.75 + 6.392 + 1.4 = 16542

6. 13.37 × 4.5 = 60165

7. 23.97 − (2.875 + 6.01) = 15085

8. 38.164 ÷ 4.7 = 812

■ Estimate an answer to each problem below. Then use your estimate to help you choose the correct answer from the choices given.

9. Three metal strips are placed side by side. How wide are they together if the individual widths are 6.95 cm, 9.1 cm, and 7.325 cm?

   (1) 14.985   (3) 20.185
   (2) 17.245   (4) 23.375

10. As a receptionist, Jade earns $6.89 per hour. How much will Jade earn in a week in which she works 41.75 hours?

   (1) $231.74   (3) $329.85
   (2) $287.66   (4) $381.08

11. Ali bought a 3.9-pound chicken on sale for $0.96 per pound. How much change will Ali get when paying with a $10 check?

   (1) $5.08   (3) $7.92
   (2) $6.26   (4) $8.34

12. Blaine bought a 29.8-pound bag of potting soil for $4.88. About how many pounds of soil did she get per dollar?

   (1) 6   (3) 10
   (2) 8   (4) 12

Rounding to a Chosen Place Value

Smaller Place Values

Once in a while, you may see a decimal fraction with more than three decimal places. In such cases, you'll need to be familiar with place values smaller than 0.001.

Decimal Place Values

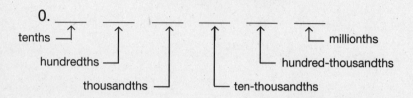

Math Facts
0.1 is one-tenth
0.01 is one-hundredth
0.001 is one-thousandth
0.0001 is one ten-thousandth
0.00001 is one hundred-thousandth
0.000001 is one-millionth

Rounding to a Chosen Place Value

As discussed on the previous two pages, rounding mixed decimals to the nearest whole number *before doing the math* is one way to estimate an answer.

Rounding is also used to simplify an exact answer *after the math is done*. Most often, an answer is rounded to a chosen place value, giving the answer fewer decimal digits.

To round a decimal fraction to a chosen place value, look at the digit to the *right* of the chosen place value.
- If the digit is 5 or more, round up.
- If the digit is less than 5, leave the digit in the chosen place value unchanged.

Rounding to the tenths place (nearest tenth)
Check the digit in the hundredths place.

0.75 ≈ 0.8	1.43 ≈ 1.4	3.489 ≈ 3.5
└ 5 or more	└ less than 5	└ 5 or more

Rounding to the hundredths place (nearest hundredth)
Check the digit in the thousandths place.

0.248 ≈ 0.25	1.271 ≈ 1.27	2.1453 ≈ 2.15
└ 5 or more	└ less than 5	└ 5 or more

Rounding to the thousandths place (nearest thousandth)
Check the digit in the ten-thousandths place.

0.9375 ≈ 0.938	3.0573 ≈ 3.057	5.21875 ≈ 5.219
└ 5 or more	└ less than 5	└ 5 or more

Practice

◼ Round each number in rows 1 and 2 to the tenths place.

1. 0.46 ≈   0.52 ≈   0.85 ≈   0.235 ≈   0.912 ≈

2. 2.35 ≈   1.83 ≈   3.901 ≈   5.052 ≈   6.750 ≈

◼ Round each number in rows 3 and 4 to the hundredths place.

3. 0.345 ≈   0.742 ≈   0.2753 ≈   0.892 ≈   0.3468 ≈

4. 3.625 ≈   4.052 ≈   8.500 ≈   6.9057 ≈   12.65805 ≈

◼ Round each number in row 5 to the thousandths place.

5. 0.3542 ≈   0.4627 ≈   2.93756 ≈   3.141593 ≈   27.00639 ≈

6. Tami, a grocery clerk, used a calculator to multiply *pounds (lb.)* by *dollars per pound ($ per lb.)* to find the selling price of each package of beef listed on the chart below. Her calculator answers are shown.

Find the selling price of the packages by rounding each calculator answer to the nearest cent (hundredths place).

	Weight (lb.)		Price ($ per lb.)	Calculator Answer		Selling Price
Hamburger	6.05	×	$1.79	10.8295	a)	_____
Round steak	3.39	×	$2.09	7.0851	b)	_____
Sirloin steak	2.80	×	$3.98	11.144	c)	_____
Rib steak	1.85	×	$4.18	7.733	d)	_____
T-bone steak	5.25	×	$4.80	25.2	e)	_____
Tenderloin steak	4.15	×	$6.79	28.1785	f)	_____
Note: A calculator does not display a dollar sign ($).						

Adding Decimals

To add decimals
- place the numbers in a column, lining up the decimal points (Remember: A whole number is understood to have a decimal point to the right of the ones digit.)
- use place-holding zeros if necessary to give all numbers the same number of decimal places (Extra zeros help keep columns in line.)
- add the columns
- place a decimal point in the answer directly below the decimal points in the problem

> **Math Tip**
>
> Write place-holding zeros at the *right* end of decimal fractions. In this way, you do not change their values.
>
> $3.5 = 3.50 = 3.500$

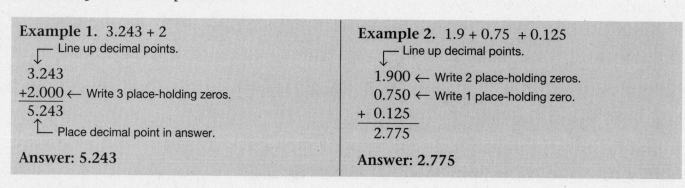

Example 1. $3.243 + 2$

Line up decimal points.

```
 3.243
+2.000  ← Write 3 place-holding zeros.
 5.243
```
Place decimal point in answer.

Answer: 5.243

Example 2. $1.9 + 0.75 + 0.125$

Line up decimal points.

```
1.900  ← Write 2 place-holding zeros.
0.750  ← Write 1 place-holding zero.
+ 0.125
 2.775
```

Answer: 2.775

Calculator Solution of Example 1
Press Keys: (C) (3) (·) (2) (4) (3) (+) (2) (=) *
Answer: (5.243)

*Note: You do not enter extra 0s when using a calculator.

Practice

■ Add. Use zeros as placeholders where necessary.

1.
```
   0.6          0.85          2.4         $0.56        $3.28
 + 0.3        + 0.62        + 1.7        + 0.49       + 1.39
```

2.
```
   0.83          2.4          7.63          9           3.875
   0.51          1.35         2             5.375       1.6
 + 0.125       + 0.96       + 1.875       + 3.65       + 5
```

■ Line up the decimal points and add. Round each answer to the tenths place.

3. $8.75 + 6.3 + 4.325$ $2.875 + 1 + 0.25$ $7 + 4.5 + 2.125$

Subtracting Decimals

To subtract decimals
- place the numbers in a column, lining up the decimal point
- use place-holding zeros if needed to give all numbers the same number of decimal places
- subtract the columns just as you subtract whole numbers
- place a decimal point in the answer directly below the decimal point in the problem

Example: 8 − 5.325

Line up the decimal points.	Use zeros as placeholders.
↓	↓↓↓
8.	8.000
− 5.325	− 5.325
	2.675

Answer: 2.675

Calculator Solution of Example
Press Keys: ⓒ ⑧ ⊖ ⑤ ⊙ ③ ② ⑤ ⊜
Answer: 2.675

Practice

■ **Subtract. Use zeros as placeholders where necessary.**

1.
0.9	0.74	5.7	13.9	$12.98
− 0.4	− 0.58	− 3.9	− 5.9	− 3.99

2.
0.875	6.8	4.56	21.9	30
− 0.35	− 0.95	− 2.275	− 8.583	− 16.735

■ **Line up the decimal points and subtract. Round each answer in row 3 to the tenths place.**

3. 2.25 − 1.9   9.85 − 4.7   18.2 − 9.75   35.5 − 25.875

■ **Round each answer in row 4 to the hundredths place.**

4. 1.25 − 0.5625   7.4 − 5.875   12 − 9.625   23.5 − 8.9375

Applying Your Skills

Solve the following decimal addition and subtraction problems.

1. At a picnic, two tables were placed end to end. The first table is 2.75 meters long, and the second table is 2.5 meters long. What is the combined length of the two tables?

2. During the first half of the baseball season, Ramon's batting average was 0.284. During the second half of the season, his average improved by 0.13. Was Ramon's batting average over 0.300 by the end of the season?

3. The area of the United States is 3.62 million square miles, while the area of Mexico is 0.76 million square miles. By how many million square miles is the United States larger than Mexico?

4. Roger lives 8.75 miles from Bay City Airport and 12.4 miles from Kingsly Field Airport. How much closer is Roger to the airport at Bay City than to the one at Kingsly Field?

5. A Danish dining table has three leaves (removable center boards), the widths of which are shown below. Without the leaves, the table is 138 cm long. What is the length of the table with all three leaves in place?

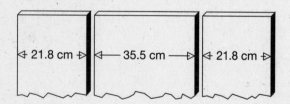

6. Cindy, a machinist, must cut a shaft so that the finished diameter (distance across) is 6.875 inches. If she starts with an 8-inch-diameter rod, how much will she need to reduce the diameter?

7. When he had the flu, Trong's temperature went from 98.6°F to 102.9°F. By how many degrees did Trong's temperature rise?

8. Before Joan had her car worked on, it got exactly 23 miles to the gallon. Now, thanks to a tune-up, it gets an additional 5.8 miles to the gallon. How many miles to the gallon does Joan's car get now?

9. When he ran the 100-meter dash, Wade finished in exactly 12 seconds. How much slower is his time than the school record of 11.375 seconds?

10. Lynn has a bracket that is exactly 6 cm long. To the nearest mm, how much longer is the bracket than the bolt pictured below?

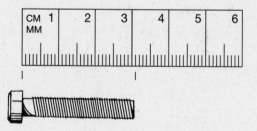

11. To pay for a $7 million convention center, the city council raised $1.24 million from the sale of bonds, $0.4 million from state economic development funds, and $2 million from private donations. How much more money does the council need to raise?

12. Carol is running in a 6.2-mile (10-km) road race. The first 1.25 miles are a slow uphill stretch, and the next 0.875 mile is a winding downhill section. The rest of the course is on level highway. How far will Carol be from the finish when she reaches the end of the downhill section?

13. To save postage, Myer is mailing several items in one box—but he must keep the total weight under 25 pounds. The box itself weighs 1.75 pounds, and the two items now in it weigh 6.25 pounds and 7.5 pounds. Which equation below tells how many more pounds (*p*) Myer can place in the box?

   (1) $p = 25 - (7.5 + 6.25 - 1.75)$
   (2) $p = (7.5 + 6.25 + 1.75) - 25$
   (3) $p = (7.5 + 6.25 - 1.75) - 25$
   (4) $p = 25 - (7.5 + 6.25 + 1.75)$
   (5) $p = 25 + (7.5 + 6.25 - 1.75)$

14. Last week, Georgia earned $4.55 as her allowance. Her sister, Cara, earned $3.85. Her brother Bill earned $2.50 less than Georgia and Cara earned together. How much did Bill earn last week?

Problems 15–16 refer to the circle graph.

**Composition of Halley's Mixed Nuts
(ingredients by weight in 1-pound mixture)**

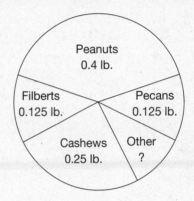

15. In each pound of Halley's Mixed Nuts, by how much do the peanuts outweigh the cashews?

16. What part of a pound is represented by the word *other* on the graph above?

For problem 17, use the chart to help you plot the trends.

17. The table below compares the value of WRT Company with that of CBA Company during the first 2 years of operation. If the present trends continue, after how many years will the two companies have the same value?

Comparative Dollar Values of WRT and CBA Companies (in millions)

Starting Value	After 1 Year	After 2 Years	After 3 Years	After 4 Years	After 5 Years
WRT Company: 2.4	3.65	4.9			
CBA Company: 4.4	5.15	5.9			

Problem Solver: **Using Drawings**

Sometimes word problems can seem confusing. We've all seen problems of this type! One technique that's very helpful is to use a drawing to organize the information. The saying "A picture is worth 1,000 words" is especially true for math problems.

> Marcella, Tony, and William carpool to QRS Electronics, where they work. All three live on 46th Street east of QRS. Marcella lives 13.9 miles farther from work than Tony. William lives 11.5 miles closer to work than Marcella. Tony lives 12.3 miles from QRS. How far does William live from QRS?

Step 1. Make a drawing, starting with the easiest fact to represent.
Tony lives 12.3 miles from QRS Electronics.

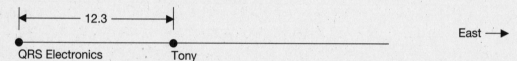

Step 2. Represent the other two facts on this drawing.
Marcella lives 13.9 miles farther from work than Tony.
William lives 11.5 miles closer to work than Marcella.

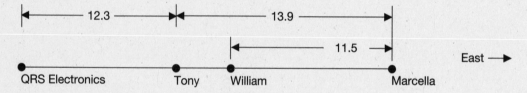

Step 3. Use the drawing to solve the problem.
   a) Determine how far Marcella lives from QRS Electronics:
      12.3 + 13.9 = 26.2 miles
   b) William lives 11.5 miles closer than Marcella:
      26.2 − 11.5 = 14.7 miles
Answer: William lives 14.7 miles from QRS Electronics.

Practice

■ **For each problem, use a drawing to organize the facts. Then solve.**

1. Karin, Shauna, and Amanda all live on Walnut Road west of the library. Karin lives 3.4 miles farther from the library than Shauna. Amanda lives 1.6 miles closer to the library than Karin. Shauna lives 5.8 miles from the library. How far does Amanda live from the library?

   Drawing

2. An airplane is flying at an altitude of 35,000 feet above sea level. A cloud, 18,500 feet below the airplane, is passing over Mount Scott, the summit of which is 9,500 feet above sea level. How far is the cloud above the summit of Mount Scott?

   Drawing

3. Stacey knows a secret and told it to Jade. During the day, Jade told 4 friends the secret. Each of these friends told 3 other friends. How many people, not counting Stacey, know the secret now?

   Drawing

4. Between 7:00 A.M. and 10:30 A.M., the temperature rose 24ºF. However, the temperature rose only 9º between 9:00 A.M. and 10:30 A.M. What was the temperature at 9:00 A.M. if the temperature at 7:00 A.M. was 47ºF?

   Drawing

5. From his home on Filmore Street, Tuan walked 1.4 miles directly east. He then turned onto 34th Street and walked 0.6 miles directly south. At that point he turned onto Harrison Street and walked 0.9 miles directly west. How far is Tuan now from home if he can get there by walking directly north to Filmore Street and then turning west to go home?

   Drawing

6. Bobbie left Seattle at 8:00 A.M. and drove south toward Eugene at a speed of 60 miles per hour. Diane left Eugene at 9:00 A.M. and drove north toward Seattle at a speed of 50 miles per hour. If Seattle is 300 miles from Eugene, how far apart were Bobbie and Diane at 10:00 A.M.? (Hint: As a first step, determine how far each woman had driven by 10:00 A.M.)

   Drawing

Multiplying Decimals

To multiply decimals
- multiply the numbers
- count the number of decimal places in the numbers being multiplied to see how many decimal places are in the answer
- count off the number of decimal places starting at the *right* of the answer; then place the decimal point

 Math Fact

The number of decimal places in a number is the number of digits to the right of the decimal point. A whole number has 0 decimal places.

Example 1

```
  3.65      2 decimal places
× 8       + 0 decimal places
 2920       2 decimal places
```
⌐ Place the decimal point so that the
answer has 2 decimal places.

Answer: 29.20 = 29.2

Example 2

```
  9.25      2 decimal places
× 3.5     + 1 decimal place
 4625       3 decimal places
+2775
32375       Place the decimal point so that the
```
answer has 3 decimal places.

Answer: 32.375

Calculator Solution of Example 2
Press Keys:
Answer: (32.375) *

*Note: A calculator automatically places the decimal point in the correct place.

Practice

▓ **Place a decimal point in each worked problem in row 1.**

1. 1.43	2 places	89	0 places	8.5	1 place	18.5	1 place
× 5	+ 0 places	× 0.03	+ 2 places	× 0.4	+ 1 place	× 0.06	+ 2 places
715	2 places	267	2 places	340	2 places	1110	3 places

▓ **Multiply. Use estimation or a calculator to check your answers.**

2. 4.5	0.23	15	$0.45	$3.29
× 5	× 6	× 0.7	× 8	× 9

3. 2.16	10.5	1.52	$4.08	0.91
× 26	× 3.4	× 0.94	× 5.5	× 7.5

■ **Line up the digits and multiply. Round each answer in row 4 to the tenths place.**

4. 3.87×5 0.56×9 1.73×0.86 0.75×21

■ **Round each answer in row 5 to the hundredths place.**

5. $\$6.48 \times 7.6$ 45.7×3.65 31.25×0.6 $\$32.20 \times 4.63$

Using Zero as a Placeholder

In some problems, it is necessary to write one or more 0s as placeholders before you can place the decimal point in the answer.

Example 3	Example 4
$\begin{array}{rl} 0.5 & \text{1 decimal place} \\ \times\ 0.03 & +\text{2 decimal places} \\ \hline \_.\_15 & \text{3 decimal places} \end{array}$	$\begin{array}{rl} 0.87 & \text{2 decimal places} \\ \times\ 0.002 & +\text{3 decimal places} \\ \hline .\_\_174 & \text{5 decimal places} \end{array}$
↑	↑ ↑
Add 1 zero. Although there are only two digits, there must be *three decimal places* in the answer.	Add 2 zeros. Although there are only three digits, there must be *five decimal places* in the answer.
Answer: 0.015	**Answer: 0.00174**
↑⎿ For practice, write a leading 0 as part of the answer.	↑⎿ Write a leading 0 as part of the answer.

Practice

■ **Multiply.**

6. $\begin{array}{r} 0.8 \\ \times\ 0.02 \\ \hline \end{array}$ $\begin{array}{r} 0.65 \\ \times\ 3.2 \\ \hline \end{array}$ $\begin{array}{r} 0.72 \\ \times\ 0.46 \\ \hline \end{array}$ $\begin{array}{r} 1.34 \\ \times\ 0.018 \\ \hline \end{array}$ $\begin{array}{r} 2.61 \\ \times\ 0.053 \\ \hline \end{array}$

7. 0.4×0.3 0.06×0.9 0.0027×8 0.25×0.016

Dividing Decimals

Dividing by a Whole Number

To divide a decimal by a whole number
- place a decimal point in the quotient directly above its position in the dividend
- use one or more 0s as placeholders in the dividend as needed

Example 1. Divide 31.2 by 6.
Step 1. Place a decimal point in the quotient.
Step 2. Divide as you do with whole numbers.

Step 1

$$6\overline{)31.2}$$

Be sure the decimal points are lined up.

Step 2

$$\begin{array}{r} 5.2 \\ 6\overline{)31.2} \\ -30 \\ \hline 12 \\ -12 \end{array}$$

Answer: 5.2

Example 2. Divide 8 into 0.184.
Step 1. Place a decimal point in the quotient.
Step 2. Since you can't divide 8 into 1, write a 0 above the 1. Now divide 8 into 18.

Step 1

$$8\overline{)0.184}$$

Write a leading 0 as part of the answer.

Step 2

$$\begin{array}{r} .023 \\ 8\overline{)0.184} \\ -16 \\ \hline 24 \\ -24 \end{array}$$

Answer: 0.023

Calculator Solution of Example 2
Press Keys: (C) (·) (1) (8) (4) (÷) (8) (=)
Answer: (0.023) *

* Note: Most calculators display a leading 0 when displaying a decimal fraction.

Practice

Divide. Use estimation or a calculator to check your answers.

1. $4\overline{)8.32}$ $5\overline{)0.755}$ $3\overline{)24.6}$ $12\overline{)0.684}$ $14\overline{)2.884}$

2. $6\overline{)0.444}$ $8\overline{)0.424}$ $6\overline{)0.774}$ $13\overline{)1.105}$ $27\overline{)0.0567}$

3. $4\overline{)\$9.16}$ $3\overline{)\$8.25}$ $5\overline{)\$12.75}$ $4\overline{)\$0.16}$ $8\overline{)\$0.24}$

Dividing by a Decimal

To divide a decimal by a decimal
- change the divisor to a whole number by moving its decimal point to the far right
- move the decimal point in the dividend an equal number of places to the right, adding place-holding zeros

Example 3. Divide: $0.03 \overline{)7.92}$

Step 1. Move the decimal point in the divisor and dividend 2 places to the right.

Step 2. Divide 3 into 792.

$$
\begin{array}{r}
264 \\
0.03 \overline{)7.92} \\
-6 \\
\hline
19 \\
-18 \\
\hline
12 \\
-12 \\
\hline
\end{array}
$$

Answer: 264

Example 4. Divide: $0.004 \overline{)1.6}$

Step 1. Move the decimal point in the divisor 3 places to the right. Add two 0s to the dividend; then move its decimal point 3 places to the right.

Step 2. Divide 4 into 1600.

$$
\begin{array}{r}
400 \\
0.004 \overline{)1.600} \\
-16 \\
\hline
000 \\
\end{array}
$$

Answer: 400

Practice

Divide.

4. $0.02 \overline{)0.46}$ $0.14 \overline{)0.42}$ $0.05 \overline{)7.485}$ $1.2 \overline{)20.64}$ $5.6 \overline{)29.12}$

5. $0.03 \overline{)1.5}$ $0.014 \overline{)30.8}$ $0.012 \overline{).144}$ $0.024 \overline{)26.4}$ $0.016 \overline{)0.208}$

To divide into a whole number, your first step is to add a decimal point to the whole number. Then use place-holding zeros and move the decimal point the appropriate number of places.

6. $0.07 \overline{)21}$ $0.03 \overline{)18}$ $0.05 \overline{)10}$ $2.5 \overline{)50}$ $1.8 \overline{)279}$

Using Place-Holding Zeros When Dividing Whole Numbers

Use place-holding zeros to
- write a remainder as a decimal fraction when dividing whole numbers
- change a proper fraction to an equivalent decimal fraction

Example 5. Divide 5 into 12 and write the remainder as a decimal fraction.	**Example 6.** Change $\frac{3}{4}$ to a decimal fraction.
Place a decimal point in the quotient.	Place a decimal point in the quotient.
$$\begin{array}{r} 2.4 \\ 5\overline{)12.0} \\ -10 \\ \hline 20 \\ -20 \end{array}$$ ← Write a decimal point and add one place-holding 0.	$$\begin{array}{r} .75 \\ 4\overline{)3.00} \\ -2.8 \\ \hline 20 \\ -20 \end{array}$$ ← Write a decimal point and add two place-holding 0s.
Answer: 2.4	**Answer: 0.75** └─ Write a leading 0 in the answer.

Practice

Divide. Add enough 0s so that each answer ends without a remainder.

7. $2\overline{)5}$ $5\overline{)24}$ $4\overline{)33}$ $4\overline{)19}$ $8\overline{)27}$

Divide to change each proper fraction to a decimal fraction with no remainder. Before dividing, reduce proper fractions to lowest terms.

8. $\frac{2}{4} =$ $\frac{6}{8} =$ $\frac{14}{20} =$ $\frac{5}{8} =$ $\frac{7}{16} =$

A **repeating decimal** occurs when you divide two numbers and continue to get a remainder with a repeating pattern. For example: $\frac{1}{3} = 0.3333333\ldots$

Two Ways to Represent a Repeating Decimal
$\frac{1}{3} = 0.3\ldots$ or $0.\overline{3}$

Each fraction below has a repeating decimal equivalent. Divide; then round each answer to the hundredths place.

9. $\frac{1}{3} \approx$ $\frac{2}{3} \approx$ $\frac{1}{6} \approx$ $\frac{5}{6} \approx$ $\frac{5}{9} \approx$

Multiplying or Dividing by 10, 100, or 1,000

To *multiply* a number by 10, 100, or 1,000, use one of these shortcuts. You may need to add one or more place-holding zeros before writing the decimal point in the product.

To multiply by 10, move the decimal point *one place* to the right.	To multiply by 100, move the decimal point *two places* to the right.	To multiply by 1,000, move the decimal point *three places* to the right
$0.6 \times 10 = 6$	$0.85 \times 100 = 85$ added 0	$0.325 \times 1,000 = 325$ added 0s
$7.5 \times 10 = 75$	$14.7 \times 100 = 1,470$	$24.6 \times 1,000 = 24,600$

To *divide* a number by 10, 100, or 1,000, use one of these shortcuts. You may need to add one or more place-holding zeros before writing the decimal point in the quotient.

To divide by 10, move the decimal point *one place* to the left.	To divide by 100, move the decimal point *two places* to the left.	To divide by 1,000, move the decimal point *three places* to the left
$0.4 \div 10 = 0.04$	$20.6 \div 100 = 0.206$	$126 \div 1,000 = 0.126$
$2.9 \div 10 = 0.29$	$5.2 \div 100 = 0.052$ added 0	$3.4 \div 1,000 = 0.0034$ added 0s

Practice

Multiply or divide as indicated.

1. $0.5 \times 10 =$ $0.86 \times 10 =$ $10 \times 3.7 =$

2. $0.75 \times 100 =$ $100 \times 2.6 =$ $15.2 \times 100 =$

3. $0.8 \times 1,000 =$ $9.25 \times 1,000 =$ $1,000 \times 25.6 =$

4. $0.7 \div 10 =$ $6.5 \div 10 =$ $12.7 \div 10 =$

5. $28.7 \div 100 =$ $3.4 \div 100 =$ $0.5 \div 100 =$

6. $125 \div 1,000 =$ $67 \div 1,000 =$ $9.8 \div 1,000 =$

Applying Your Skills

Solve the following decimal multiplication and division word problems.

1. At Vern's Market, Ellie bought a 6.1-lb. package of hamburger selling at $1.29 per lb. Which of the following is the best estimate of what Ellie will pay for this purchase?

   (1) $6.00
   (2) $6.90
   (3) $7.80
   (4) $8.50
   (5) $9.00

2. While traveling through Oregon, Annick drove 301.6 miles on 14.5 gallons of gas. Knowing this, figure out Annick's average mileage (miles per gallon) on this trip. Round your answer to the nearest mile per gallon.

3. Yann packed 30 cans of smoked salmon in a large shipping container. Each can of fish weighs 2.875 pounds, and the container itself weighs 23.25 pounds. What is the total weight of the packed container? Round your answer to the nearest pound.

4. In the metric system, road distance is measured in kilometers. A kilometer (km) is a little shorter than a mile: 1 kilometer is equal to 0.62 mile. Giving your answer in miles, how much shorter is 50 km than 50 miles?

   |← ———————— 50 miles ———————— →|

   |← ——— 50 kilometers ——— →|← ? miles —→|

5. Kevin worked 19.6 hours of overtime last month, for which he was paid $243.98. Which expression below gives the best estimate of Kevin's overtime hourly pay rate?

   (1) $200 × 10
   (2) $250 × 20
   (3) $200 ÷ 20
   (4) $250 ÷ 19
   (5) $250 ÷ 20

6. Jill earns "time and three-quarters" for each hour of overtime she works. Her overtime pay rate is found by multiplying her regular pay rate by 1.75. What is Jill's overtime rate if her regular hourly rate is $6.80?

7. On Friday, Josue sold 4 pickup loads of topsoil at his garden shop. What is the average weight of these 4 loads? Round your answer to the nearest hundredth ton.

   Load #1: 0.874 ton   Load #3: 0.73 ton
   Load #2: 1.05 ton   Load #4: 0.5 ton

8. As shown below, a machinist placed 4 equal-size washers on a 1.125-inch bolt. The length of the uncovered part of the bolt is 0.875 inch. What is the exact width of each of the washers?

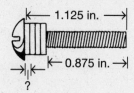

9. A box arrived at Kim's Import Food Store weighing 24.5 kilograms. The box contained 24 jars of pickled vegetables. When empty, the box weighs 1.6 kilograms. Which equation below tells how many kilograms (k) each jar weighs?

(1) $k = (24.5 - 1.6) \div 24$
(2) $k = (24.5 - 24) \div 1.6$
(3) $k = (24.5 - 1.6) \times 24$
(4) $k = (24.5 - 24) \times 1.6$
(5) $k = (24.5 + 1.6) \div 24$

10. A *digital scale* shows weight as a mixed decimal number of pounds. On most digital scales, weight is rounded to the nearest hundredth pound. Many markets, delis, and other shops use a digital scale to weigh a customer's purchase. What total price should the scale below show?

Scale	
Total Price	**?**
4.73 lb.	$3.89
Weight	$ per lb.

11. Nadine needs to drill a hole that will allow a 0.325-inch-diameter wire to pass through. She wants the wire to fit as tightly as possible. Which of the three drill bits shown below should she use?

Bit #	Diameter
#1	$\frac{5}{16}$ inch
#2	$\frac{11}{32}$ inch
#3	$\frac{3}{8}$ inch

12. A 3-foot-wide bookshelf contains 20 books. How much room is left on the bookshelf if 10 of the books are 0.75 inch wide and 10 of the books are 1.125 inches wide?

13. At his office supplies store, Jackson got a package from UPS that weighed 53 pounds. In the package are 1,000 pens. If the box itself and the packing material have a total weight of 2.5 pounds, what is the approximate weight of each pen?

14. In Fremont, the cost of electric power is $0.0956 per kilowatt-hour. If, during the month of January, the Smith family used 3,000 kilowatt-hours of electric power, how much was their electric bill for that month? (Hint: Multiply; then round your answer to the nearest cent.)

■ **Problem 15 is based on the drawing below.**

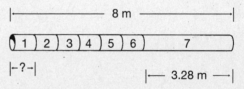

15. Shaun is going to cut the 8-meter-long pipe into 7 pieces as shown above. First, he will cut off a piece 3.28 meters long; then, he will cut the remaining piece into 6 pieces of equal length. Which expression below gives the length of each of these shorter pieces?

(1) $\frac{8 - 3.28}{7}$

(2) $\frac{7 - 3.28}{7}$

(3) $\frac{8 - 3.28}{6}$

(4) $\frac{7 - 3.28}{6}$

(5) None of the above

Problem Solver: Working Backward

In some problems, you are given an end value and asked to find a previous value. For problems of this type, you'll often find that *working backward* is the best approach. Here are two examples.

Example 1

After getting up, Jerry takes 45 minutes to eat and get ready for work. He then drives 10 minutes to Frank's, where he meets Frank, Orin, and Jesse. Together they carpool to work, a drive that takes 35 minutes. They always allow 30 minutes of extra time to have coffee at a restaurant before arriving at work.

a) What time must the men leave Frank's if they need to be at work by 9:00 A.M.?

Work backward, using the stated facts to find the time of each event.

Time they arrive at coffee shop: 9:00 A.M. − 30 minutes = 8:30 A.M.
Time they leave Frank's: 8:30 A.M. − 35 minutes = 7:55 A.M.

Answer: The men must leave Frank's by 7:55 A.M.

b) What time must Jerry get up if he wants to be at work by 9:00 A.M.?

Work backward from the answer in part **a**.

Time Jerry leaves for Frank's: 7:55 A.M. − 10 minutes = 7:45 A.M.
Time Jerry must get up: 7:45 A.M. − 45 minutes = 7:00 A.M.

Answer: Jerry must get up by 7:00 A.M.

Example 2

Rent at Village Apartments has risen drastically between 1980 and today. Between 1980 and 1990, rent doubled. Between 1990 and 1992, rent increased another $60 per month. Luckily, today's rent of $545 per month is only $35 higher than it was in 1992. How much was monthly rent at Village Apartments in 1980?

Work backward, using the stated facts to figure the monthly rent for each given year.
a) Rent today: $545
b) Rent in 1992: $545 − $35 = $510
c) Rent in 1990: $510 − $60 = $450
d) Rent in 1980: $450 ÷ 2 = $225

Answer: Rent in 1980 was $225 per month.

Practice

▪ **Working backward, solve each problem.**

1. Mrs. Johnson kept 12 of the cookies she made and placed them in a container for her family. She took the rest of the cookies to her daughter's class. The teacher, Mr. Swenson, gave half of these cookies to his class and the other half to Mrs. Blake's class. Mrs. Blake had 32 cookies to share with her students.

   a) How many cookies did Mr. Swenson and Mrs. Blake share?

   b) How many cookies did Mrs. Johnson bake?

2. After she wakes up, it takes Jillian 1 hour and 30 minutes to eat, get her daughter off to school, and get herself ready to leave for work. She then has a 10-minute walk to the bus station and a 15-minute bus ride to the stop close to her work. From this stop, it is a 5-minute walk to work.

   a) To be at work by 9:30 A.M., what time must Jillian catch the bus?

   b) What time must Jillian get out of bed if she wants to be to work by 9:30 A.M.?

3. Jackson used $\frac{1}{2}$ of a full box of nails while working on the upstairs of the new house. Loretta, who was working on the main floor, took $\frac{1}{2}$ of the nails left in the box. When Tom went to get nails, he found only $1\frac{1}{2}$ pounds of nails remaining.

   a) How many pounds of nails were in the box just before Loretta took some out?

   b) How many pounds of nails were in the full box to begin with?

4. Tuesday's low temperature was 12 degrees higher than Monday's low. Wednesday's low of 42°F was 9 degrees colder than Tuesday's. Use these facts to determine Monday's low temperature reading.

5. The starting salary for technicians at Jenson Electronics has gone up substantially between 1960 and today. Between 1960 and 1965, the starting salary increased by $1.50 per hour. Between 1965 and 1990, the starting salary doubled. Between 1990 and today, starting salary increased by another $1.75 per hour. If today's starting salary is $11.65, what was the starting salary in 1960?

Data Highlight: Reading a Line Graph

A **line graph** gets its name from a line or lines that it uses to connect graphed data points. Numerical values are read along numbered scales called **axes** that make up the sides of the graph. The value of each graphed point is read as two numbers, one taken from each axis.

The graph below represents the average oxygen consumption rate of healthy, nonathletic adults of average build.

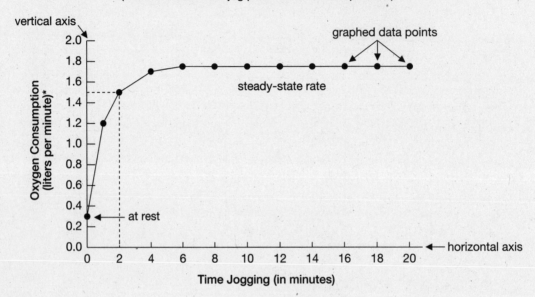

Average Oxygen Consumption Rate, Measured While Jogging
(measured at a slow-jog pace of 12 minutes per mile)

Examples

A. What is the approximate oxygen consumption rate of an average person at rest?

B. What is the approximate oxygen consumption rate of an average person after 2 minutes of jogging?

Answers

A. A person at rest would have spent 0 minutes jogging.
Find 0 on the *horizontal* axis (axis running left to right). Now read the data point indicated on the *vertical axis* (axis running up and down).
Answer: A person at rest consumes about 0.3 liters of oxygen per minute.

B. *Step 1.* Locate the graphed data point that is directly above the 2 on the horizontal axis.
Step 2. Read the value on the vertical axis that is directly to the left of this data point.
Answer: After 2 minutes of jogging, a person consumes about 1.5 liters of oxygen per minute.

*A liter is a metric capacity unit that is slightly larger than one quart.

Practice

▪ Problems 1–2 refer to the line graph on page 124.

1. Determine the approximate oxygen consumption rate of an average person after 4 minutes of jogging.

2. What is the approximate ratio of an average person's steady-state rate of oxygen consumption to his or her rate while at rest?

▪ Problems 3–6 refer to the line graph below.

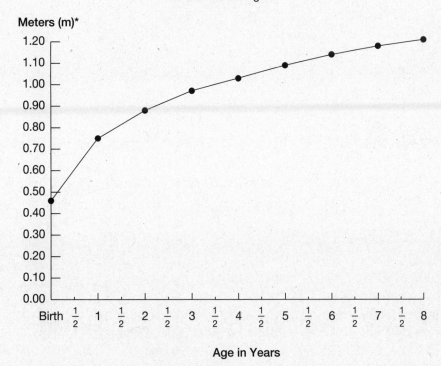

Height Growth Chart: Average-Size Girl
From Birth to Age 8

Meters (m)*

Age in Years

*When a length is written in meters, the first two decimal places are centimeters (cm).
For example, 0.30 m = 30 cm, and 1.10 m = 110 cm.

3. What is the approximate length (height) of an average-size baby girl at birth:

in m? _____   in cm? _____

4. Approximately how much height (in cm) does a young girl gain between her first and third birthdays?

5. Between which two birthdays does an average-size girl reach the height of 1 meter?

6. During which year does a baby girl's height increase most rapidly?

Test Readiness Checkup

Circle each correct answer. Check your answers on page 245; then correct any errors.

1. Anna bought a 2.94-pound salmon for $3.89 a pound. What computation can you do to get a quick estimate of Anna's change if she pays with a $20 bill?

   (1) $20 − ($4 × 2)   (4) $20 − ($3 × 2)
   (2) $20 + ($4 × 3)   (5) $20 + ($3 × 2)
   (3) $20 − ($4 × 3)

2. Jeff delivers furniture for Mill's Furniture Store. On Saturday, Jeff drove 4.6 miles to the warehouse to pick up a bedroom set. He then drove 18.9 miles to the Malloys' to deliver it. He then drove 13 miles back to Mill's Furniture. How many miles did Jeff drive for this delivery?

   (1) 17.42   (4) 31.4
   (2) 21.3   (5) 36.5
   (3) 28

3. Jacques's new skis are 1.87 meters long. How much shorter are these skis than the 2-meter skis he used last year?

   (1) 13 cm   (4) 103 cm
   (2) 23 cm   (5) 213 cm
   (3) 83 cm

Problem 4 refers to the following drawings.

1.1875"   0.875"

4. What is the difference in length of the two nails pictured above?

   (1) 0.15"   (4) 0.875"
   (2) 0.3125"   (5) 1.1"
   (3) 0.625"

5. In the metric system, small weights are measured in grams: 28.4 g ≈ 1 oz. Knowing this, determine the weight in grams of an 8.75-ounce can of tuna.

   (1) 3.2   (4) 248.5
   (2) 48.8   (5) 386.2
   (3) 126.4

6. A batting average is figured by dividing the number of hits a player gets by the number of times he or she bats. Last year Carmine was at bat 80 times and got 25 hits. To the nearest thousandth, what was Carmine's batting average?

   (1) 0.264   (4) 0.358
   (2) 0.287   (5) 0.408
   (3) 0.313

7. Joni paid $16.15 for a 7.5-foot piece of oak molding. To the nearest cent, how much did Joni pay per foot?

   (1) $1.25   (4) $8.95
   (2) $2.15   (5) $23.65
   (3) $4.45

Problem 8 refers to the drawing below.

brick   mortar   0.5"

|←————————— 38" —————————→|

8. Which expression below can be used to find the length of each brick shown above?

   (1) $\frac{38" - 0.5"}{4}$ (4) $\frac{38" - (4 \times 0.5")}{4}$

   (2) $\frac{38" - 0.5"}{3}$ (5) $\frac{38" + (3 \times 0.5")}{4}$

   (3) $\frac{38" - (3 \times 0.5")}{4}$

9. Which fraction below is a repeating decimal?

(1) $\frac{1}{8}$　　　　(4) $\frac{1}{3}$

(2) $\frac{1}{5}$　　　　(5) $\frac{1}{2}$

(3) $\frac{1}{4}$

10. One hundred tiles were laid in a row. Each tile was 9 inches wide, and a space of 0.25 inches was left between tiles. To the nearest foot, how long is this row of tiles?

(1) 64 feet　　(4) 92 feet
(2) 77 feet　　(5) 99 feet
(3) 83 feet

11. At the Paint Place, interior wall paint is on sale for $11.79 a gallon. Interior enamel is on sale for $6.19 a quart. Which expression represents the best estimate of the cost of buying 4 gallons of interior wall paint and 3 quarts of interior enamel?

(1) 3($6) + 4($12)
(2) 3($12) + 4($6)
(3) (3 + 4)($6 + $12)
(4) 3($5) + 4($11)
(5) 3($6) + 4($11)

12. Ellie is planning a turkey dinner for Sunday. She wants to start eating at 5:00 P.M. She plans to cook the turkey for 3.5 hours, and she wants to allow a total of 20 minutes for cooling and carving time before dinner starts. What time should Ellie put the turkey in the oven?

(1) 12:20 P.M.
(2) 12:40 P.M.
(3) 1:10 P.M.
(4) 1:40 P.M.
(5) 2:00 P.M.

13. What is the speed in kilometers per hour of a car traveling 55 mph? (1 mi ≈ 1.6 km)

(1) 58　　　　(4) 80
(2) 64　　　　(5) 88
(3) 76

14. Zoe's calculator displays 8 digits, but it does not display unnecessary zeros. If Zoe divides 14 by 3 on her calculator, what answer will be displayed?

(1) 0.46　　　　(4) 4.7777777
(2) 4.6　　　　(5) 46.666666
(3) 4.6666666

■ **Problems 15–16 refer to the line graph.**

Temperatures in New York City

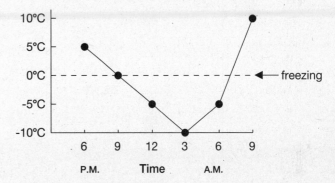

15. What was the temperature in New York City at 9:00 P.M.?

(1) 10ºC　　　　(4) −5ºC
(2) 5ºC　　　　(5) −10ºC
(3) 0ºC

16. About how many degrees Celsius did the temperature increase between 6:00 A.M. and 9:00 A.M.?

(1) 5　　　　(4) 12
(2) 7　　　　(5) 15
(3) 10

UNIT 6 Percents

Identifying Numbers in Percent Problems

Percent problems involve three important numbers: **percent**, **whole**, and **part**.

Example 1	Example 2
25% of $100 is $25. percent whole part	15 is 50% of 30. part percent whole

Percent problems ask you to find one of these numbers when you know the other two.

* Find the *part* when you know the percent and whole.
  What is 18% of $50?
  percent = 18%   whole= $50   part = ?

* Find the *percent* when you know the whole and part.
  What percent of 75 is 15?
  whole = 75   part = 15   percent = ?

* Find the *whole* when you know the percent and part.
  If 20% of a number is 14, what is the number?
  percent = 20%   part = 14   whole = ?

> **Math Tip**
>
> In percent problems, the *whole* always follows the word *of*.

Practice

▧ **Identify the *percent*, *whole*, and *part* in each problem below.**

1. 20% of 90 is 18.

   a) percent =____
   b) whole =____
   c) part =____

2. $4.80 is 60% of $8.00.

   a) percent =____
   b) whole =____
   c) part =____

▧ **Circle the number of what you are asked to find in each problem below.**

3. 30% of a number is 60. What is this number? *You have to find the* ____ .

   (1) percent   (2) whole   (3) part

4. 15 is what percent of 75? *You have to find the* ____ .

   (1) percent   (2) whole   (3) part

The Percent Circle

Each type of percent problem is solved by multiplication or division. To remember whether to multiply or to divide, we'll use a memory aid called the **percent circle.** The symbols *P*, %, and *W* are used on this circle.

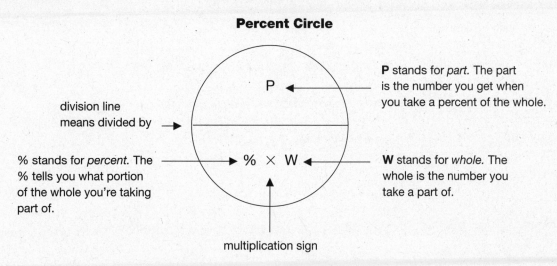

Percent Circle

P stands for *part*. The part is the number you get when you take a percent of the whole.

division line means divided by →

% stands for *percent*. The % tells you what portion of the whole you're taking part of.

W stands for *whole*. The whole is the number you take a part of.

multiplication sign

Example 1	Example 2
Identify P, %, and W in the statement below. 30% of 120 is 36.	Identify P, %, and W in the statement below. 40 is 50% of 80.
Answer:　P　= 36 　　　　　% = 30% 　　　　　W = 120	Answer:　P　= 40 　　　　　% = 50% 　　　　　W = 80

Practice

▊ Identify P, %, and W in each statement below.

1. 42 is 10% of 420.

   a) P = _____
   b) % = _____
   c) W = _____

2. 65% of 200 is 130.

   a) P = _____
   b) % = _____
   c) W = _____

3. $36 is 20% of $180.

   a) P = _____
   b) % = _____
   c) W = _____

▊ Circle the symbol of what you are asked to find.

4. What is 75% of 240?

   P　%　W

5. 7 is what percent of 35?

   P　%　W

6. What percent of 150 is 50?

   P　%　W

Using the Percent Circle

To use the percent circle
- cover the symbol of the number you're trying to find
- do the math indicated by the uncovered symbols

Example 1. Finding *part* of the whole.

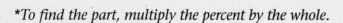

If 15% of your $1,200 monthly paycheck is used for your car payment, how much is this payment?

Step 1. Cover P (the part)—the number you're trying to find.

Step 2. Read the uncovered symbols: % × W

$P = \% \times W$

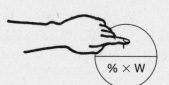

To find the part, multiply the percent by the whole.

Example 2. Finding what *percent* a part is of a whole.

If $300 of your $1,200 monthly paycheck is used to pay rent, what percent is your rent payment?

Step 1. Cover % (the percent)—the number you're trying to find.

Step 2. Read the uncovered symbols: $\frac{P}{W}$

$\% = \frac{P}{W}$ (means P ÷ W)

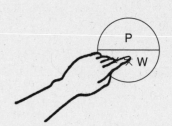

To find the percent, divide the part by the whole.

Example 3. Finding a *whole* when a part and a percent are given.

Suppose you buy a new 27" color TV set and make a 10% down payment of $52. What is the price of the set?

Step 1. Cover W (the whole)—the number you're trying to find.

Step 2. Read the uncovered symbols: $\frac{P}{\%}$

$W = \frac{P}{\%}$ (means P ÷ %)

To find the whole, divide the part by the percent.

Practice

■ **Fill in the percent circle. Then complete the three sentences below.**

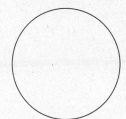

1. To find the part, _____

2. To find the percent, _____

3. To find the whole, _____

■ **Circle the symbol of what you're asked to find in each problem. Then decide whether you solve the problem by multiplication or division and put a check next to your choice.**
Do not solve these problems.

4. Leona pays 25% of her monthly income for rent. How much is Leona's rent payment if her monthly salary is $1,180?

   a) P % W

   b) ____ multiplication   ____ division

5. Seven of the 24 children in Mrs. Altman's first-grade class were home on Tuesday with colds. What percent of the class was sick that day?

   a) P % W

   b) ____ multiplication   ____ division

6. Suppose you make a down payment of $38 on a new sewing machine. If this down payment is 15% of the selling price, how much are you being charged for the machine?

   a) P % W

   b) ____ multiplication   ____ division

7. If you are charged a sales tax of $0.75 on a new $14.99 shirt, what percent is the sales tax of the cost of the shirt?

   a) P % W

   b) ____ multiplication   ____ division

8. According to the new report, 64% of the voters voted in favor of the school budget. If 28,680 people voted, how many voted in favor of the budget?

   a) P % W

   b) ____ multiplication   ____ division

Changing Percents to Fractions or Decimals

As shown on the percent circle, many percent problems are solved by multiplying or dividing by a percent. However, to multiply or divide by a percent, you must first change the percent to a fraction or a decimal. Then you multiply or divide in the usual way.

On these next two pages, we want to strengthen your skill (first learned on pages 62 and 63) in changing a percent to either a fraction or a decimal.

Changing a Percent to a Fraction

To change a percent to a fraction
- write the percent as a fraction with a denominator of 100
- reduce the fraction if possible

Change 40% to a fraction.

Step 1. Write 40% as 40 over 100. $40\% = \frac{40}{100}$

Step 2. Reduce $\frac{40}{100}$ by dividing both top and bottom numbers by 20. $\frac{40 \div 20}{100 \div 20} = \frac{2}{5}$

Answer: $\frac{2}{5}$

Math Tip

The percents $33\frac{1}{3}\%$ and $66\frac{2}{3}\%$ occur so often in real-life problems that you may want to memorize the following facts.

$33\frac{1}{3}\% = \frac{1}{3}$ $66\frac{2}{3}\% = \frac{2}{3}$

Changing Percents to Decimals

To change a percent to a decimal
- move the decimal point two places to the left, adding one or two zeros if necessary
  Remember: The decimal point of a whole number is understood to be at the right of the number, even though it may not be written.
- drop the percent sign
- drop any unnecessary zeros

Percent	Move Decimal Point Two Places Left	Decimal
25%	25.	0.25 — Write a leading 0.
60%	60. — Add 1 zero.	0.60 = 0.6 — Drop the unnecessary 0.
7%	07. — Add 2 zeros.	0.07
.5%	00.5	0.005

Practice

■ **Change each percent to an equivalent fraction.**

1. 30%   50%   85% $66\frac{2}{3}$% $33\frac{1}{3}$%

2. 1%   5%   9%   8%   6%

■ **Change each percent to an equivalent decimal.**

3. 25%   50%   75%   90%   7%

4. 5.5%   9.9%   8.5%   0.6%   0.1%

■ **Solve each problem below.**

5. At a Memorial Day sale, The Men's Shop reduced the prices of the items listed at right. Write an equivalent fraction for each listed discount.

Sale Item	Discount	Fraction Off
a) Shirts	25%	_____
b) Sweaters	$33\frac{1}{3}$%	_____
c) Jackets	10%	_____
d) Gloves	60%	_____
e) Boots	$66\frac{2}{3}$%	_____
f) Scarves	75%	_____

6. When working with money, you can think of percent as meaning "cents per dollar." For example, 3% can be thought of as 3¢ per dollar. Why? Because 3% = $\frac{3}{100}$, and 3¢ = $\frac{3}{100}$ of $1.

   Complete the table at right, showing how many cents per dollar each percent represents.

Percent	Cents per Dollar
a) 5%	_____
b) 7%	_____
c) 12%	_____
d) 25%	_____
e) 50%	_____
f) 75%	_____
g) 90%	_____

Finding the Part

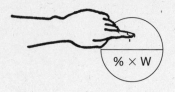

Percent Circle: To find part of a whole, multiply the percent by the whole.

How to Do It: Change the percent to a decimal or a fraction; then multiply.

P = % × W

Example: An $80 radio is on sale for "25% off." How much can you save by buying the radio on sale?

You save 25% of $80.

Method 1*	Method 2*
Step 1. Change 25% to a decimal. 25% = 0.25	*Step 1.* Change 25% to a fraction. $25\% = \frac{25}{100} = \frac{1}{4}$
Step 2. Multiply $80 by 0.25. $\begin{array}{r} \$80 \\ \times\ 0.25 \\ \hline 400 \\ 1600 \\ \hline \$20.00 \end{array}$	*Step 2.* Multiply $80 by $\frac{1}{4}$. $\frac{\$80}{1} \times \frac{1}{4} = \frac{\$80}{4}$ = $20 **Answer: $20**
Answer: $20	

*For most problems, you may find method 1 easier. But when the percent is $33\frac{1}{3}\%$ (= $\frac{1}{3}$) or $66\frac{2}{3}\%$ (= $\frac{2}{3}$), method 2 is certainly easier!

Calculator Solution of Example
Press Keys: (C) (8) (0) (×) (2) (5) (%) *   *On some calculators, you must press = after %.
Answer: (20.)

Practice

Find each part.

1. 25% of 68   15% of 200   50% of 128 $33\frac{1}{3}$ % of 45

2. 6% of 75 $66\frac{2}{3}$ % of $126   9.5% of 300   5.5% of $50

3. $8\frac{1}{2}$ % of $70 $6\frac{1}{2}$ % of $200   250% of 96   475% of $2,000

(Hint: $8\frac{1}{2}$% = 8.5%.)   (Hint: 250% = 2.5.)

Solve each problem below.

4. In a state with a 6% sales tax, how much tax must be paid on a sweater that normally sells for $32?

5. Savings Mart is offering a 25% discount on all sport shirts in stock. What discount will be given on a sport shirt that normally sells for $30?

6. Randi's property tax bill is increasing by 3% this coming year. She now pays $890 a year in property tax.

   a) By how much is her property tax going to increase?

   b) How much will she pay next year?

7. This morning, Lyle learned that he is getting a $5\frac{1}{2}$ % raise. Before the raise, his salary was $320 per week.

   a) By how much is Lyle's weekly salary going to increase?

   b) What is Lyle's new weekly salary?

8. The newspaper reported that 8.5% of the 11,800 accidents in the state this year involved cars 15 years old or older. To the nearest 100, how many accidents involved these older cars?

9. AGB Computer Company announced a $33\frac{1}{3}$ % increase in sales this year. By how much did sales increase this year if last year's sales were $24 million?

Problems 10–11 are based on the graph below.

Lyford Family Budget
(Monthly Income: $1,500)

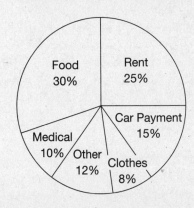

10. How much do the Lyfords spend on food each month?

11. What is the Lyfords' monthly rent payment?

Finding the Percent

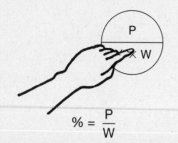

$$\% = \frac{P}{W}$$

Percent Circle: To find the percent, divide the part by the whole.

How to Do It: Write a fraction: $\frac{P}{W}$.

Reduce this fraction; then multiply by 100%.

Example: In Alicia's first-grade class, 6 of the 24 students are foreign-born. What percent of Alicia's class is foreign-born?

Step 1. Write the fraction 6 (P) over 24 (W). $\frac{P}{W} = \frac{6}{24} = \frac{1}{4}$
Reduce this fraction.

Step 2. Multiply $\frac{1}{4}$ by 100%. $\frac{1}{\cancel{4}} \times \frac{\cancel{100}\%}{1} = 25\%$

Answer: 25%

Calculator Solution of Example
Press Keys: Ⓒ ⑥ ÷ ② ④ %* *On some calculators, you must press = after %.
Answer: 25.

Practice

Solve each problem below. Remember to reduce the $\frac{P}{W}$ fraction before multiplying by 100%.

1. What percent of 40 is 8?

2. $5 is what percent of $20?

3. What percent of 16 is 12?

4. 9 gallons is what percent of 27 gallons?

5. What percent of 8 pounds is 3 pounds?

6. 18 inches is what percent of 1 yard?
   (1 yard = 36 inches)

Practice

■ **Solve each problem below.**

7. On his math test, Walker got 48 questions correct out of 64. What percent of the test questions did Walker answer correctly?

8. When Hansen Real Estate sold Jim's house, it charged him a commission of $4,500. If the sale price of the house was $75,000, what percent real estate commission was he charged?

9. Out of her monthly gross pay of $1,200, Yoshi's employer withholds $180 for federal tax. Determine the percent of Yoshi's salary that is withheld for this tax.

10. The Sport House pays $12 for each basketball it buys. It then sells each ball for $20 and makes an $8 profit. Knowing this, figure out what percent markup (profit) is used by The Sport House.
(Hint: The whole is the original price.)

11. Three years after Beth bought a new Honda for $14,500, its value decreased to $8,700.

   a) By how much did the value of the Honda drop in three years?

   b) What percent depreciation (value decrease) is this over the three-year period?

12. During the last year, the price of a $350 dishwasher increased to $400. To the nearest percent, what percent inflation (price increase) is shown by this change in price?

13. During the fall, the weather report was correct only 26 times out of the 92 days that it predicted the next day's weather. Which expression below gives the percent of times it made *incorrect* predictions?

   (1) $\frac{26}{92} \times 100\%$

   (2) $\frac{92 + 26}{92} \times 100\%$

   (3) $\frac{92 - 26}{92} \times 100\%$

   (4) $\frac{92}{92 + 26} \times 100\%$

   (5) $\frac{92}{92 - 26} \times 100\%$

■ **Problems 14–15 are based on the table below.**

Sale Prices at Gregory's		
Item	**Original Price**	**Sale Price**
Sport Coat	$80.00	$64.00
Wool Pants	60.00	45.00
Wool Sweater	48.00	32.00
Dress Shoes	99.00	33.00

14. How much money is saved by buying each of the items listed above at the sale price?

   a) Sport coat: ____   c) Wool sweater: ____

   b) Wool pants: ____   d) Dress shoes: ____

15. What percent of the original price do you save by buying each listed item at the sale price?

   a) Sport coat: ____   c) Wool sweater: ____

   b) Wool pants: ____   d) Dress shoes: ____

Finding the Whole

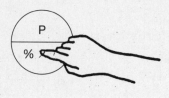

$$W = \frac{P}{\%}$$

Percent Circle: To find the whole, divide the part by the percent.

How to Do It: Change the percent to a decimal or a fraction; then divide.

Example: 30% of the people who applied this week for work at the new electronics plant were hired. If 60 people were hired, how many people submitted applications?

In other words, 60 is 30% of what number?

Method 1	Method 2
Step 1. Change 30% to a decimal.	*Step 1.* Change 30% to a fraction.
$30\% = 0.30 = 0.3$	$30\% = \frac{30}{100} = \frac{3}{10}$
Step 2. Divide 60 by 0.3.	*Step 2.* Divide 60 by $\frac{3}{10}$.
$\overset{200}{0.3\overline{)60.0}}$	$60 \div \frac{3}{10} = \frac{60}{1} \times \frac{10}{3}$
	$= \frac{600}{3} = 200$
Answer: 200 people applied	**Answer: 200 people applied**

Calculator Solution of Example
Press Keys: 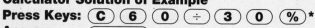 ⓒ ⑥ ⓪ ÷ ③ ⓪ ⑨ *   *On some calculators, you must press = after %.
Answer: 200.

Practice

■ **Using either method 1 or method 2, solve these problems.**

1. 15% of what number is 45?

2. $36 is 20% of what amount?

3. 27 tons is 30% of what weight?

4. $66\frac{2}{3}$ % of what number is 150?

(Hint: $66\frac{2}{3}$ % = $\frac{2}{3}$.)

Practice

Solve each problem below.

5. To pass his math test, Keith needs to get 60% of the questions correct. If Keith needs 42 correct answers to pass, how many questions are on the test?

6. Fifteen percent of Joyce's monthly income goes toward her car payment. If her car payment is $165, how much does Joyce make each month?

7. When he bought a new TV, Maurice made a 25% down payment of $87.50. What was the total price of the TV?

8. Dee was charged $4.50 during June on the unpaid balance on her Visa card. Her Visa company charges a 1.5% finance charge each month on her unpaid balance. How much was Dee's unpaid balance during June?

9. Jordan bought a shirt marked "25% off." Jordan paid $27 for the shirt.

   a) What *percent* of the original price of the shirt did Jordan pay?

   b) Using your answer from part a, figure out the shirt's original price.

10. During the month of May, 40% of the babies born at St. Paul Hospital were boys. During that month, 90 girls were born at St. Paul.

    a) What *percent* of the babies born at St. Paul during May were girls?

    b) Using your answer from part a, figure out how many babies in all were born in St. Paul during May.

11. At a clearance sale, Kelli bought a blouse marked "70% off" for $19.89.

    a) What percent of the original price did Kelli pay for the blouse?

    b) Which expression below gives the best estimate of the original price of this blouse?

       (1) $\frac{\$20}{0.3}$

       (2) $\frac{\$20}{0.4}$

       (3) $\frac{\$20}{0.5}$

       (4) $\frac{\$20}{0.6}$

       (5) $\frac{\$20}{0.7}$

Problem 12 is based on the circle graph.

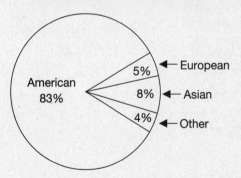

**Student Enrollment Makeup
Blaine Community College**

12. a) If 800 European students attend Blaine Community College, what is the total student enrollment at Blaine?

    b) Using your answer from part a, determine how many Asian students attend Blaine.

    c) What is the ratio of Asian students at Blaine to European students?

Calculator Spotlight: Increasing or Decreasing an Amount

Many percent problems involve increasing or decreasing a whole by a part. For problems of this type, a calculator greatly simplifies the steps involved. On these next two pages, you'll see just how useful a calculator can be in problem solving!

Suppose you want to find the purchase price of a car in a state where there is a sales tax.

Without a calculator, you perform two steps:
- First, you multiply to find the amount of sales tax (the part).
- Second, you add the sales tax to the selling price (the whole).

Using a calculator, you combine these two steps into a single step.

Example: In a state with a 5% sales tax, what is the purchase price of a car listed at $8,500?

Step 1. Identify % and W.
% = 5%   W = $8,500

Step 2. On your calculator, add
$8,500 and 5% of $8,500
by pressing keys as follows:

	Press Keys	Display Reads
Clear the display.	(C)	0.
• Enter 8,500.	(8)(5)(0)(0)	8500.
• Press (+)	(+)	8500.
• Enter 5, the number of percent.	(5)	5.
• Press (%).*	(%)	8925.

Answer: $8,925

*On most calculators, pressing (%) completes this calculation. On some calculators, though, you must press (%) and then (=).

Calculator Facts

When you press (8)(5)(0)(0)(+)(5)(%)*, your calculator automatically adds 8,500 and 5% of 8,500.

If you want to subtract a percent from a whole (as you do in discount problems) press (−) instead of (+).

Example: Pressing (2)(0)(0)(−)(6)(%)* subtracts 6% of 200 from 200 and gives the answer 188.

*On some calculators, you press (=) to complete this calculation.

Practice

■ **Do the problems below with paper and pencil. Then do them with a calculator. Which way is easier for you? Do you see why a calculator is such a valuable tool?**

Rate Increase: new amount = original amount + amount of increase

1. By changing jobs, Dave increased his monthly salary by 7%. What does Dave earn now if his previous job paid him $1,230 per month?

2. This year, the Jacobsens' average weekly food bill has been $86. However, experts predict a 4.5% rise in food prices for next year. If they're right, how much (to the nearest dollar) can the Jacobsens expect to pay in weekly food bills next year?

Rate decrease: new amount = original amount − amount of decrease

3. By turning their thermostat down to 68 degrees during the winter months, the Mondales hope to save 6% on their monthly heating bills. If they succeed, how much will the Mondales' heating bills average this winter if last winter they averaged $182.50?

4. To lose weight, Gregory has been advised to cut his daily calorie intake by 30%. Before starting this diet, Gregory consumed about 2,700 calories each day. While on the diet, what target level (to the nearest 100 calories per day) should Gregory try to attain?

Markup: selling price = store's cost + markup

5. Central Hardware places a 30% markup on each item it sells. If Central pays $13.50 for StrongArm Hammers, what price will Central charge its own customers for these hammers?

6. At Hermann's Men's Store, Hermann pays $59.00 for Paris Nights sweaters. Hermann then adds a 35% markup to his cost. What does Hermann charge his customers for these sweaters?

Discount: sale price = original price − amount of discount

7. During the July 4th sale, Valley Appliances is marking down its appliances by 20%. What will be the sale price of a washer that normally sells for $289?

8. Salem Motors is offering a 15% discount on any new car in stock. Interested in a good deal, the Carmino family is looking at a new compact car that has a sticker price of $9,800. To the nearest hundred dollars, what will be the discounted price?

Life Skill: Understanding Simple Interest

Interest is money that you earn (or pay) for the use of money.
- If you deposit money in a savings account, *the bank pays you interest* for the use of your money.
- If you borrow money, *you pay the lender interest* for the use of its money.

Simple interest is interest earned (or paid) on the amount of principal—the amount that is deposited or borrowed.

To compute simple interest, you use the simple interest formula:

In words: Interest equals *Principal* times *Rate* times *Time*
In symbols: I = PRT (which means P × R × T)

Interest (I) =	Principal (P)* ×	Rate (R) ×	Time (T)
Expressed in dollars	Expressed in dollars	Expressed as a percent	Expressed in years

*Be sure not to confuse the use of the letter *P* in the simple-interest formula with its use in the percent circle. In I = PRT, *P* stands for *principal*. In the percent circle, *P* stands for *part*.

Example 1

How much interest is earned on $600 deposited for 3 years in a savings account that pays 5% simple interest?

Step 1. Identify P, R, and T.

P = $600   R = 5% = 0.05   T = 3 yr.

Step 2. Replace the letters in PRT with the values given in Step 1.

I = PRT = $600 × 0.05 × 3

$$\begin{array}{r} \$600 \\ \times\ 0.05 \\ \hline \$30.00 \end{array} \longrightarrow \begin{array}{r} \$30 \\ \times\ 3 \\ \hline \$90 \end{array}$$

Answer: $90

Note: You can also write $5\% = \frac{5}{100}$
In this case, you would multiply as follows:

$$I = \$\overset{6}{\cancel{600}} \times \frac{5}{\cancel{100}} \times 3 = \$90$$

Example 2

How much interest would you pay on $500 loaned to you at a simple-interest rate of 8.5% for 2 years?

Step 1. Identify P, R, and T.

P = $500   R = 8.5% = 0.085   T = 2 yr.

Step 2. Replace the letters in PRT with the values given in Step 1.

I = PRT = $500 × 0.085 × 2

$$\begin{array}{r} \$500 \\ \times\ 0.085 \\ \hline 2500 \\ 40000 \\ \hline \$42500 \end{array} \longrightarrow \begin{array}{r} \$42.50 \\ \times\ 2 \\ \hline \$85.00 \end{array}$$

Answer: $85.00

Practice

▪ Use the simple-interest formula to solve the problems.

1. How much would you earn on a deposit of $2,500 in 2 years if you were paid a simple-interest rate of 6%?

2. Suppose you deposit $800 in a savings account that earns 4.5% simple interest.

   a) How much interest will you earn in 2 years?

   b) What will be the total in your account at the end of the 2 years?
   (balance due = principal + interest)

3. The loan rates charged by States Bank are shown below. Suppose you borrow $8,000 to buy a used car. The bank agrees to let you pay the entire principal plus interest at the end of 3 years. What total amount will you owe the bank at the end of 3 years?
   (balance owed = principal + interest)

States Bank Simple-Interest Loans	
New Car	9.9%
Used Car	10.5%
Boat	12.6%
Personal	14.8%

Repayment Schedules

When you borrow money from a bank or credit union, you will usually repay it according to a **repayment schedule**. A repayment schedule lists interest rates, loan amounts, and monthly payments.

▪ Problems 4–5 are based on the repayment schedule shown below.

MONTHLY REPAYMENT SCHEDULE				
		Repayment Period		
Rate	Loan Amount	24 months	36 months	48 months
10%	$5,000	$231	$161	$127
	$7,500	$346	$242	$191
12%	$5,000	$235	$166	$131
	$7,500	$353	$249	$197

Math Tip

In a repayment schedule, an interest charge is included as part of each monthly payment.

4. Suppose you take out a $7,500 loan at an interest rate of 10% and plan to repay it in 24 months.

   a) What will be your monthly payment?

   b) How much money will you pay to the credit union during the 24 months?

   c) How much total interest will you pay for this loan?
   (total interest = total payments − $7,500)

5. Suppose you take out a $5,000 loan at an interest rate of 12% and plan to repay it in 48 months.

   a) What will be your monthly payment?

   b) How much total interest will you pay for this loan?

   c) How much interest can you save on this $5,000 loan by paying it off in 36 months rather than in 48 months?

143

Computing Interest for Part of a Year

Interest is usually listed as a yearly rate. However, many deposits or loans involve parts of a year.
When using the simple-interest formula for part of a year, write the time as either a fraction or a decimal—depending on which one makes the multiplication easier.

Example 1

How much interest is earned on a $600 deposit if you are paid an interest rate of 6% and you leave your money in the bank for 8 months?

Step 1. Identify P, R, and T.

$$P = \$600 \quad R = 6\% = \frac{6}{100} \quad T = \frac{8}{12} = \frac{2}{3}$$

8 mo. → $\frac{8}{12}$ ← 12 mo. = 1 yr.

Step 2. Multiply.

$$PRT = \$600 \times \frac{6}{100} \times \frac{2}{3}$$

$$= \$\cancel{600} \times \frac{\overset{6}{\cancel{6}}}{\underset{1}{\cancel{100}}} \times \frac{\overset{2}{2}}{\underset{1}{\cancel{3}}}$$

$$= \$24$$

Answer: $24

Example 2

How much interest will you pay on a loan of $700 borrowed for 1.5 years at a 14% interest rate?

Step 1. Identify P, R, and T.

$$P = \$700 \quad R = 14\% \quad T = 1.5$$

Step 2. Multiply.

$$PRT = \$700 \times 0.14 \times 1.5$$

$$= \$98 \times 1.5$$

$$= \$147$$

Answer: $147

Practice

Solve.

1. Write each time period as a proper fraction. Reduce if possible.

   4 months   6 months   9 months   11 months

2. Write each time period as a decimal fraction. Round long answers to the hundredths place.

   3 months   4 months   9 months   10 months

3. Write each time period as indicated.

   1 year and 3 months   2 years and 4 months   2 years and 9 months

   improper fraction: _____   decimal: _____   improper fraction: _____

4. How much interest will you earn on a $400 deposit placed in a savings account for 9 months if you're paid at a simple-interest rate of 5%?

5. Suppose you take out a $650 loan. How much interest will you owe at the end of 8 months if you are charged an interest rate of 10%?

6. At a bank that pays 4% simple interest, how much can you earn on a deposit of $900 in 18 months?

7. Suppose you put $1,000 in a savings account that pays 5% simple interest. What will be your balance in the account after 15 months?
(balance = principal + interest)

8. You borrow $1,500 and are charged 14% simple interest. What total amount will you owe the lender after 1 year and 9 months?
(balance owed = principal + interest)

9. How much will it cost you to repay a $700 loan after 15 months if you are paying a simple-interest rate of 12%?

Charge-Card Interest Rates

Millions of people have a *charge card*. Many of these cards charge 1.5% *per month* interest on any unpaid balance. Often called "a small monthly finance charge," a 1.5% monthly rate is actually equal to an annual percentage rate (APR) of 18%!

10. Suppose you have a Visa card that charges 1.5% per month on any unpaid balance.

    a) Calculate the monthly finance charge you would owe at the end of each month listed at right.
    (finance charge = unpaid balance × 1.5%)

    b) Suppose you charged $400 on your Visa and made no monthly payments for 1 year. How much interest would you be charged for the year on this $400? (yearly interest = principal × APR × 1)

Month	Unpaid Balance	Finance Charge
May	$80.00	_____
June	$120.00	_____
July	$200.00	_____
August	$350.00	_____

11. Suppose you are going to make a $600 purchase using a charge card. You plan to pay off the balance quickly and pay interest for only 1 month. Which of the two Visa cards below would be less expensive for you for this purchase?

National Visa: Charges 1.5% per month

Diamond Visa: Charges 1% per month plus a $2 monthly service charge

Problem Solver: Estimating with Easy Percents

In many percent problems (including many test questions), you do not need to compute an exact answer. An estimate will be close enough.

Be Familiar with These "Everyday Percent" Rules

The following rules come from changing everyday percents to fractions.

- 10% is the same as $\frac{1}{10}$ of a number. To find 10% of a number, divide by 10.
  Example: Find 10% of 10. **Solution:** $10 \div 10 = 1$

- 25% is the same as $\frac{1}{4}$ of a number. To find 25% of a number, divide by 4.
  Example: Find 25% of 48. **Solution:** $48 \div 4 = 12$

- $33\frac{1}{3}$% is the same as $\frac{1}{3}$ of a number. To find $33\frac{1}{3}$% of a number, divide by 3.
  Example: Find $33\frac{1}{3}$% of 30. **Solution:** $30 \div 3 = 10$

- 50% is the same as $\frac{1}{2}$ of a number. To find 50% of a number, divide by 2.
  Example: Find 50% of 82. **Solution:** $82 \div 2 = 41$

Work with Compatible Numbers

A good way to estimate is to round to everyday percents and compatible numbers whenever possible. *Compatible numbers* are numbers that you can easily multiply or divide.

Example 1. Finding the part.
What is 24% of 79?
Think: $24\% \approx 25\%$ *and* $79 \approx 80$
Rule: To find 25% of 80, divide 80 by 4.

$80 \div 4 = 20$

Estimate: 20

Example 2. Finding the part.
What is 11% of 412?
Think: $11\% \approx 10\%$ *and* $412 \approx 410$
Rule: To find 10% of 410, divide 410 by 10.

$410 \div 10 = 41$

Estimate: 41

Example 3. Finding the percent.
8 is what percent of 31?

Think: 8 is compatible with 32 ($32 \div 8 = 4$).

Write: $\frac{8}{31} \approx \frac{8}{32} = \frac{1}{4}$ and $\frac{1}{4} = 25\%$

Estimate: 25%

Example 4. Finding the whole.
48% of a number is 92. What is the number?

Think: $48\% \approx 50\%$ *and* $92 \approx 90$

Write: $92 \div 48\% \approx 90 \div 50\%$

$= 90 \div \frac{1}{2} = \frac{90}{1} \times \frac{2}{1} = 180$
Estimate: 180

Practice

■ **Use estimation in each problem below. As a first step, decide whether you are asked to find the *part*, the *percent*, or the *whole*.**

1. Joyce determined that 23% of her monthly income goes to the rent payment. If her monthly income is $1,189, about how much is her rent payment?

2. When he bought the new gas range, Blake paid $50 down, or 9.5% of the sale price. What was the approximate sale price of the range?

3. Mari left a tip of 70¢ for a meal that cost $3.45. Approximately what percent tip did she leave?

4. According to a recent poll, $33\frac{1}{3}$% of the people polled said that they approved of the city's plan to expand the library. If 904 people were polled, *about* how many were in favor of expansion?

5. Starting next week, Josh will receive a raise of 5.2%. He now earns $7.00 per hour. Which expression below most accurately describes the amount of Josh's raise?
(Hint: 5.2% ≈ 5% = $\frac{1}{2}$ of 10%.)

   (1) a little less than 35¢ per hour
   (2) a little more than 35¢ per hour
   (3) a little less than 70¢ per hour
   (4) a little more than 70¢ per hour
   (5) a little less than $1.05 per hour

6. 67% of the employees at Anson & Sons Electronics are men. If Anson has 609 employees, how many of these employees are *women*?
(Hint: 67% ≈ $66\frac{2}{3}$% = $\frac{2}{3}$.)

7. Twenty-three percent of Stephen's monthly check goes to pay rent. If Stephen pays $345 per month for rent, how much is his monthly check?

8. The company Velma works for pays 75% of each employee's health care costs. Each employee pays $36 a month to pay the rest of the health care coverage.

   a) How much does the company pay each month for health care coverage for each employee?

   b) How much does the company pay each year for each employee for health care coverage?

9. At a restaurant, Sherry leaves a tip of about 15%. To do this, she estimates 10% of the meal cost. Then she takes half of this (5%) and adds that amount to her 10% estimate. Which expression below is the best estimate of the tip Sherry would leave for a $14.75 meal?

   (1) ($15 ÷ 10) + ($15 ÷ 5)
   (2) ($15 ÷ 10) − $\frac{1}{2}$($15 ÷ 10)
   (3) ($15 × 10) + $\frac{1}{2}$($15 × 5)
   (4) ($15 ÷ 10) + $\frac{1}{2}$($15 ÷ 10)
   (5) ($15 ÷ 10) − ($15 ÷ 5)

Changing Fractions and Decimals to Percents

Once in a while, you need to be able to change either a fraction or a decimal to a percent. We'll study these skills on the next two pages.

Changing a Fraction to a Percent

To change a fraction to a percent, multiply the fraction by 100%.

Example 1	Example 2	Example 3
Change $\frac{3}{4}$ to a percent.	Change $\frac{7}{8}$ to a percent.	Change $\frac{2}{3}$ to a percent.
$\frac{3}{4} \times \frac{100\%}{1} = 75\%$	$\frac{7}{8} \times \frac{100\%}{1} = \frac{175\%}{2} = 87.5\%$	$\frac{2}{3} \times \frac{100\%}{1} = \frac{200\%}{3} = 66\frac{2}{3}\%$
Answer: 75%	Answer: 87.5% or $87\frac{1}{2}\%$	Answer: $66\frac{2}{3}\%$

Changing a Decimal to a Percent

To change a decimal to a percent
- move the decimal point two places to the right, adding a zero or two if necessary
- drop the decimal point if the percent is a whole number
- add a percent sign

> **Math Tip**
>
> You can think of changing a decimal to a percent as multiplying the decimal by 100%: multiply by 100 and add a percent sign. To multiply by 100, you move the decimal point two places to the right.

Decimal	Move Decimal Point Two Places to the Right	Percent
0.2	0.20 ← Add a 0.	20% ← Drop the decimal point.
0.75	0.75	75%
0.375	0.375	37.5%
1.25	1.25	125%
0.05	0.05	5% ← Drop the unnecessary 0s.
3	3.00 ← Add 2 0s.	300%

Practice

■ **Change each fraction below to a percent.**

1. $\frac{1}{4}$ $\frac{2}{5}$ $\frac{3}{8}$ $\frac{7}{10}$ $\frac{1}{3}$ $\frac{5}{6}$

■ **Change each decimal below to a percent.**

2. 0.5   0.3   0.25   0.45   0.875

3. 0.275   0.06   1.75   4.5   6

4. Below is a partially completed chart of the most commonly used percents, decimals, and fractions. Complete this chart.

Percent	Decimal	Fraction	Percent	Decimal	Fraction
	0.1				$\frac{3}{5}$
20%			$66\frac{2}{3}\%$		
	0.25			0.7	
		$\frac{3}{10}$			$\frac{3}{4}$
	$0.33\frac{1}{3}$		80%		
40%					$\frac{9}{10}$
		$\frac{1}{2}$		1	

5. At a Christmas sale, Jacob's Furniture advertised "All Chairs $\frac{1}{3}$ Off." Express this price reduction as a percent.

6. Emmett Construction completed $\frac{17}{20}$ of the apartment complex by April.

   a) By April, what percent of the complex was completed?

   b) By April, what percent of the complex was not yet finished?

7. Citizen's Bank pays $0.045 per year for each dollar placed in a savings account earning simple interest. What percent savings rate is Citizen's Bank paying?

8. 0.625 of the students in Briggs High School are women.

   a) What percent of the students in Briggs High School are women?

   b) What percent of the students in Briggs High School are men?

Data Highlight: **Reading a Bar Graph**

A **bar graph** gets its name from the bars that it uses to show data. These bars may be drawn vertically (up and down) or horizontally (across). You read a value for each bar by finding the number on the axis that is directly across from the end of the bar.

Usually, a bar graph is used to show a quick comparison of values. Exact values are often hard to obtain from a bar graph.

The bar graph below shows the average percent of fat found in several common foods.

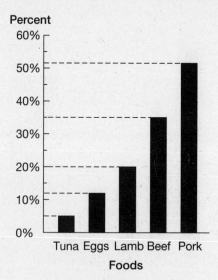

Percent of Fat in Certain Foods

Examples

A. What is the average percent of fat found in beef?

B. What is the approximate ratio of the amount of fat found in pork to the amount found in tuna?

Answers

A. *Step 1.* Locate the top of the bar labeled *Beef.*

   Step 2. Read the value on the vertical axis that is directly to the left of the top of this bar.

   Answer: about 35%

B. *Step 1.* Determine the approximate percent of fat found in pork and tuna: pork ≈ 50% and tuna ≈ 5%.

   Step 2. Determine the ratio $\frac{\text{pork}}{\text{tuna}} \approx \frac{50\%}{5\%} = \frac{10}{1}$

   Answer: about 10 to 1

Practice

▪ **Problems 1–3 refer to the bar graph on page 150.**

1. Which of the listed foods contains the highest percent of fat?

2. Of the listed foods, which contain less than 30% fat?

3. (a) What is the approximate percent of fat found in eggs?

   (b) A large egg weighs about 2 ounces. Knowing this, determine approximately how much fat is contained in a serving of 2 large eggs. Express your answer to the nearest 10th ounce.

▪ **Problems 4–9 refer to the bar graph below.**

Water Content of Selected Common Foods

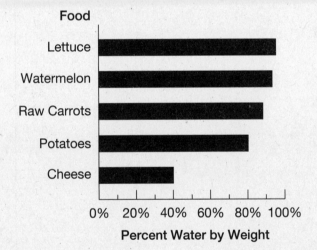

Percent Water by Weight

4. Approximately what percent of the weight of carrots is made up of water?

5. Which of the listed foods contain more than 90% water by weight?

6. To the nearest ounce, how much water is contained in a pound of carrots? Remember: 1 pound = 16 ounces.

7. How much more water is contained in 8 ounces of potatoes than in 8 ounces of cheese? Express your answer to the nearest ounce.

8. What is the approximate ratio of the percent of water in potatoes to the percent of water in cheese?

9. When you pay $0.21 per pound for watermelon, about how much are you paying per pound for water?

Test Readiness Checkup

■ **Circle each correct answer. Check your answers on page 247; then correct any errors.**

1. A magazine reported that 35% of the people who attended the rally were women. If 133 women were there, which expression could you use to find how many people attended in all?
   - (1) 133 ÷ 0.35
   - (2) 0.35 ÷ 133
   - (3) 133 ÷ 35
   - (4) 0.35 × 133
   - (5) 133 × 35

2. Which of the following has (have) the same value as 40%?
   A. 4   B. 0.4   C. $\frac{2}{5}$ D. $\frac{4}{5}$
   - (1) A and C only
   - (2) B and C only
   - (3) A and D only
   - (4) B and D only
   - (5) B only

3. Fortrell Lumber Company announced that it would rehire $66\frac{2}{3}$% of the 291 workers it laid off last year. How many workers are going to be rehired?
   - (1) 67
   - (2) 97
   - (3) 138
   - (4) 194
   - (5) 224

4. Including the sales tax, what is the total cost of the toaster shown below?

   5% Sales Tax on All Items
   - (1) $25.20
   - (2) $26.80
   - (3) $29.00
   - (4) $32.40
   - (5) $44.00

$24.00

5. For selling 25 sets of Children's Fairy Tales, Lynn was paid a $200 commission. If her total sales receipts are $1,250, what percent commission does Lynn earn?
   - (1) 8%
   - (2) 10%
   - (3) 12%
   - (4) 14%
   - (5) 16%

6. When she bought a stove, Ming made a 20% down payment of $82.50. What is the purchase price of Ming's new stove?
   - (1) $252.50
   - (2) $384.60
   - (3) $412.50
   - (4) $644.80
   - (5) $825.00

7. Clara placed $650 in a savings account that pays 4% simple interest. How much interest will Clara earn from this account if she leaves her money in for 18 months?
   - (1) $23
   - (2) $29
   - (3) $35
   - (4) $39
   - (5) $42

8. The Clothes House prices all skirts at $33\frac{1}{3}$% more than it pays for them. Which expression below shows how much customers are charged for a skirt that The Clothes House buys for $36?
   - (1) $36 − ($\frac{1}{3}$ × $36)
   - (2) $36 + ($\frac{1}{3}$ × $36)
   - (3) $36 − ($36 ÷ $\frac{1}{3}$)
   - (4) $36 + ($36 ÷ $\frac{1}{3}$)
   - (5) None of the above

9. A camera that sold for $290 last year is on sale for $240 this year. Which expression below can be used to find the percent of *price decrease* of this camera during the past year?

(1) $\frac{\$240}{\$290} \times 100\%$

(2) $\frac{\$290 - \$240}{\$240} \times 100\%$

(3) $\frac{\$290 + \$240}{\$240} \times 100\%$

(4) $\frac{\$290 - \$240}{\$290} \times 100\%$

(5) $\frac{\$290 + \$240}{\$290} \times 100\%$

12. Carla borrowed $2,960 from Lane Credit Union at a simple-interest rate of 10.25%. Which expression below gives the best estimate of the total amount Carla will owe Lane at the end of 1 year and 11 months?

(1) $\$3,000 \times \frac{1}{10} \times 2$

(2) $\$3,000 - (\$3,000 \times \frac{1}{10} \times 2)$

(3) $\$3,000 + (\$3,000 \times \frac{1}{10} \times 2)$

(4) $\$3,000 + (\$3,000 \times \frac{1}{100} \times 2)$

(5) $\$3,000 + (\$3,000 \times 10 \times 2)$

■ **Problems 10–11 refer to the graph below.**

Average Car Trade-in Value

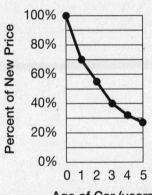

10. What is the approximate trade-in value after 1 year of a car that cost $13,800 when new?

(1) $8,540
(2) $9,660
(3) $10,850
(4) $11,320
(5) $12,400

11. Approximately what was the original purchase price of a car that has a trade-in value of $7,500 when it is 3 years old?

(1) $12,575
(2) $14,250
(3) $15,725
(4) $16,840
(5) $18,750

■ **Problems 13–14 refer to the graph below.**

Voter Turnout Rate in Benson County (recent presidential election years)

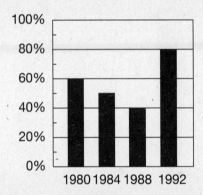

13. If Benson County had 68,400 registered voters in 1984, how many of these people did *not* vote in the 1984 election?

(1) 21,080
(2) 27,360
(3) 34,200
(4) 41,040
(5) 47,800

14. In 1992, 66,400 Benson County residents voted in the presidential election. How many total people were registered to vote in Benson County that year?

(1) 53,120
(2) 68,000
(3) 74,850
(4) 78,000
(5) 83,000

UNIT 7 Data Analysis and Probability

Locating Data

In this chapter, we'll discuss several special topics concerning **data**—a collection of factual information.
- **Data analysis** deals with interpreting data and drawing logical conclusions from it.
- **Probability** deals with our ability to make predictions based on laws of chance.

Now we'll look at two ways to estimate the value of data points that are not specifically given in a table or on a graph.

Interpolation

To **interpolate** is to *estimate* the value of a data point that lies *between* two given values.

Example 1

Referring to the table at right, determine Jesse Baker's approximate weight at age 27 months.

Noticing that 27 lies halfway between 24 and 30 months, ask, "What weight is halfway between 22 and 26 pounds?"

The answer, **24 pounds**, is your best estimate of Jesse's weight at age 27 months.

You interpolate to estimate a value that lies within the *range* of your given data. Here, the range in ages is from 18 to 36 months, and the range in weight is from 18 to 30 pounds.

Weight Growth Chart of Jesse Baker

Age (months)	Weight (pounds)
18	18
24	22
27 --------------------- ?
30	26
36	30

Extrapolation

To **extrapolate** is to *estimate* the value of a data point that lies *outside the range* of your given data.

Example 2

Refer again to the table above. About what will Jesse weigh at age 42 months?

To estimate Jesse's weight at 42 months, notice that his weight pattern up to now is an increase of 4 pounds every 6 months.

Ask yourself, "If this pattern continues, what will Jesse weigh in 6 more months—at age 42 months?" To estimate, add 4 to 30.

The answer, **34 pounds**, is your best estimate of Jesse's weight at 42 months of age.

Finding the Pattern

Age (months)	Weight (pounds)	
18	18	
		+ 4 lb.
24	22	
		+ 4 lb.
30	26	
		+ 4 lb.
36	30	
		+ 4 lb.
42	?	

Practice

▨ Problem 1 refers to the following table.

Recommended Weight of Average-Build Adults (in pounds)		
Height	Women	Men
5'2"	115	124
5'4"	122	133
5'6"	129	142
5'8"	136	151
5'10"	144	159
6'		167

1. **a)** Estimate the recommended weight of a 5'9" tall man of average build.

   b) Estimate the recommended weight of a 6' tall woman of average build.

▨ Problem 2 refers to the following line graph.

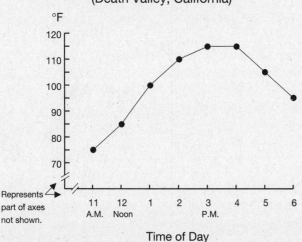

Temperature Record for July 8
(Death Valley, California)

2. **a)** What was the approximate Death Valley temperature at 12:30 P.M. on July 8?

   b) If this temperature pattern continued, what did the temperature in Death Valley drop to by 8:00 P.M.?

▨ Problem 3 refers to the following picture graph.

Monthly Sales Figures of Delmont Shoe Co.

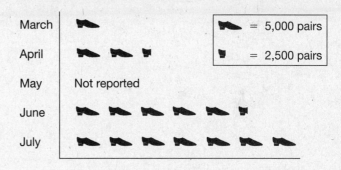

3. **a)** From the sales pattern shown above, estimate the sales figures for the month of May.

   b) Assuming the sales pattern continues, about how many shoes can Delmont expect to sell in August?

▨ Problem 4 refers to the following bar graph.

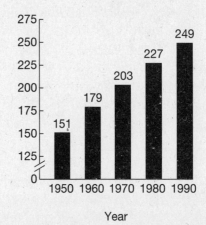

Population of the United States
(in millions)

4. **a)** Estimate what the population of the United States was in 1985.

   b) If the present growth rate continues, what will be the *approximate* population of the United States in the year 2010?

Using More than One Data Source

Sometimes you may need to use more than one data source to determine needed information.

Example

Using the information given below, determine the amount of protein in a 6-ounce serving of Millie's Famous Sandwich Spread.

**Ingredients of Millie's
Famous Sandwich Spread**
(percent by weight)

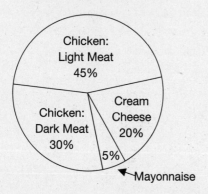

Nutrition Information		
	Protein (grams per ounce)	Fat (grams per ounce)
Chicken: Light Meat	9	1
Chicken: Dark Meat	8	2
Cream Cheese	2	10
Mayonnaise	0	11

Step 1. Determine how many ounces of each ingredient are in a 6-ounce serving. Use information given on the circle graph for this step.

ounces of ingredient = percent of ingredient × 6 ounces

Chicken: light meat = 45% × 6 = 2.7 ounces
Chicken: dark meat = 30% × 6 = 1.8 ounces
Cream cheese = 20% × 6 = 1.2 ounces
Mayonnaise = 5% × 6 = 0.3 ounce

Step 2. Compute the amount of protein in the quantity of each ingredient determined in step 1. For step 2, use information given in the table.

amount of protein = protein per ounce × number of ounces

Chicken: light meat = 9 × 2.7 = 24.3 grams
Chicken: dark meat = 8 × 1.8 = 14.4 grams
Cream cheese = 2 × 1.2 = 2.4 grams
Mayonnaise = 0 × 0.3 = 0 grams

Answer: Total grams of protein = 24.3 + 14.4 + 2.4 + 0 = 41.1 ≈ 41 grams

A 6-ounce serving of Millie's Famous Sandwich Spread contains about **41 grams** of protein.

Practice

■ **Use the circle graph and table on page 156 to solve problem 1.**

1. Fill in the table at right as you answer the following questions about Millie's Famous Sandwich Spread.

   a) How many ounces of each ingredient are in a 4-ounce serving?

   b) How many grams of fat of each ingredient are in a 4-ounce serving?

   c) How many total grams of fat are in a 4-ounce serving?

Millie's Famous Sandwich Spread (4-ounce serving)		
Ingredient	**Ounces**	**Fat (g)**
Chicken: Light Meat	_____	_____
Chicken: Dark Meat	_____	_____
Cream Cheese	_____	_____
Mayonnaise	_____	_____
Total Grams of Fat:		_____

■ **Use the bar graph and line graph below for problem 2.**

Physiological Data taken on Gregory Lin—an average-size adult male

A. Oxygen Usage Rate

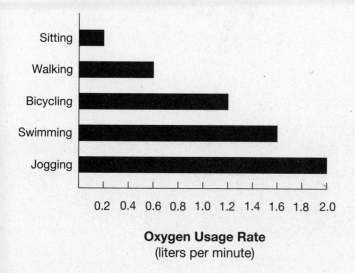

Oxygen Usage Rate
(liters per minute)

B. Calorie Usage Rate

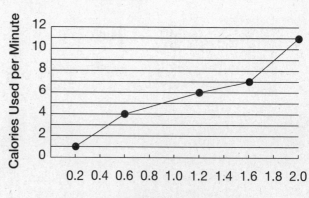

Oxygen Usage Rate
(liters per minute)

2. Fill in the table at right as you answer the questions below.

   a) What is Gregory Lin's oxygen usage rate (in liters per minute) for each of the listed activities? (See Graph A.)

   b) About how many calories per minute does Gregory Lin use for each activity? (Knowing his oxygen usage rate, use Graph B to determine calorie use.)

Activity	Oxygen Rate (Graph A)	Calorie Rate (Graph B)
Sitting	_____	_____
Walking	_____	_____
Bicycling	_____	_____
Swimming	_____	_____
Jogging	_____	_____

Use the following bar graph and table for problem 3.

Workweek: April 3–7

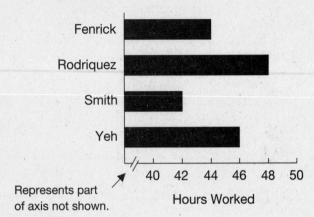

Employee

Fenrick
Rodriquez
Smith
Yeh

40 42 44 46 48 50

Represents part
of axis not shown.

Hours Worked

Hourly Pay Rate of Employees		
Employee	Regular	Overtime
Fenrick	$7.00	$10.50
Rodriquez	$7.00	$10.50
Smith	$7.00	$10.50
Yeh	$7.00	$10.50

3. Write your answers to the following questions in the table at right.

a) How much regular pay did each employee earn during the week shown?

b) How much overtime pay did each employee earn during the week shown?

Employee	Regular Pay (1st 40 hours)	Overtime Pay (Hours over 40)
Fenrick		
Rodriquez		
Smith		
Yeh		

Use the picture graph and table below for problem 4.

May Sales Report of 4 Items
(Each symbol represents 50 items)

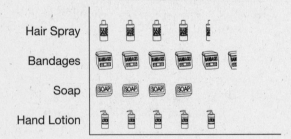

Hair Spray
Bandages
Soap
Hand Lotion

Family Market Price List		
Item	Retail Price	Profit
Hair Spray	$2.98	$1.05
Bandages	$1.99	$0.75
Soap	$0.89	$0.35
Hand Lotion	$5.77	$2.03

4. a) Determine the *approximate* total retail sales amount that Family Market received during May from the sale of bandages.

b) Determine the *approximate* total profit that Family Market made during May from the sale of hand lotion.

Drawing Conclusions from Data

To **draw a conclusion** from data is to express an idea that is logically connected to the data. Tests often ask you to determine which of several conclusions is best supported by data given in a table or graph.

Median Earnings of Full-Time Workers			
Year	Median Earnings Women	Men	Women's Earnings as a Percent of Men's Earnings
1960	$3,257	$5,368	60.7%
1970	$5,323	$8,966	59.4%
1975	$7,504	$12,758	58.8%
1980	$11,197	$18,612	60.2%
1985	$15,624	$24,195	64.6%
1989	$18,778	$27,430	68.0%

Statistical Abstract of the United States: 1992

Example

Which of the following statements is a conclusion that can be supported by data in the table?

(1) Between 1960 and 1989, more women than men worked as secretaries and assistants, and these jobs paid less than the managerial jobs held mainly by men.

(2) In 1989, the median salary of full-time working women was more than 30% less than the median salary of full-time working men.

(3) Today, most jobs pay men and women the same salary for the same work performed.

Statement 1 is true but *is not supported* by data in the table. The table does not deal with specific job categories.

Statement 2 is true and *is supported* by data in the table.

Statement 3 is false and *is not supported* by the data given. The table does not deal with work conditions after 1989.

Answer: Only **statement 2** is supported by data in the table.

Note: Although a statement is true, it may not be supported by the data given. You must carefully distinguish between conclusions that logically follow from given data and conclusions you know to be true from your own experience or from other sources of information.

Practice

◼ **Problem 1 is based on the following bar graph.**

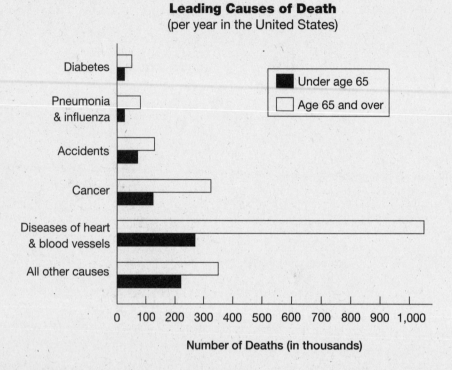

Leading Causes of Death
(per year in the United States)

Number of Deaths (in thousands)

1. Which of the following statements is (are) supported by data on the bar graph? (More than one choice is possible.)
   (1) A heart transplant is not an option in every case of severe heart disease.
   (2) Less money is spent on diabetes research than for any other cause.
   (3) For people over age 65, cancer is the leading cause of death.
   (4) For people under age 65, heart and blood vessel disease is the leading cause of death.

◼ **Problem 2 is based on the circle graph at right.**

2. Which of the following statements is (are) supported by data on the circle graph? (More than one choice is possible.)
   (1) The cost of obtaining energy from coal is about the same as obtaining energy from natural gas.
   (2) The United States uses about twice as much energy from nuclear power as energy from hydroelectric power.
   (3) The United States is the world's leading user of petroleum products (oil, gasoline, diesel fuel, jet fuel, etc.).
   (4) Less than half of the energy used in the United States comes from petroleum.

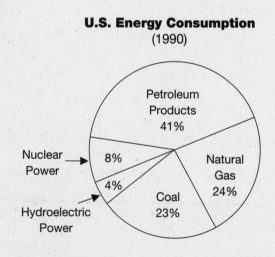

U.S. Energy Consumption
(1990)

Problem 3 is based on the following bar graph and table.

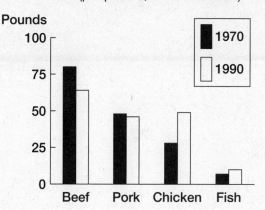

United States Meat Consumption
(per person, 1970 and 1990)

Nutritional Data for Selected Meats
(6-ounce serving)

	Protein (g)	Fat (g)
Beef		
Hamburger	42	34
Steak	48	26
Pork		
Lean Pork	32	42
Ham	36	38
Chicken		
Dark Meat	48	11
Light Meat	54	9
Fish		
Salmon	34	10
Tuna	36	8

3. Which of the following statements is (are) supported by information given on the bar graph and table above? (More than one choice is possible.)

   (1) Between 1970 and 1990, Americans increased the amount of chicken and fish they ate but decreased the amount of beef and pork.
   (2) Between 1970 and 1990, the cost of beef and pork increased while the cost of chicken and fish decreased.
   (3) Between 1970 and 1990, Americans increased the amount of fat they obtained from eating the listed meats.
   (4) Between 1970 and 1990, Americans decreased the amount of fat they obtained from eating the listed meats.

Problem 4 is based on the following circle graphs.

Current American Diet

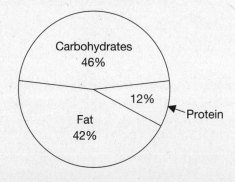

Recommended American Diet

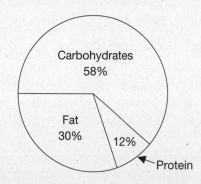

4. Which of the following suggestions is the best summary of the recommended dietary changes shown on the circle graphs?

   (1) Eat fewer carbohydrates and more fat.
   (2) Eat less protein and more carbohydrates.
   (3) Eat less fat and more protein.
   (4) Eat less fat and more carbohydrates.

Understanding Probability

Probability is the study of **chance**—the likelihood of an event happening. The word *chance* indicates our lack of control over what actually happens.

The spinner circle at right is divided into 4 equal-size sections: red, blue, green, and yellow. Assuming the spinner doesn't stop on a line, it is equally likely to stop in any of these 4 sections. We say, "Where the spinner stops is left to chance."

In the study of probability, the outcome (result) we're interested in is called a **favorable outcome**. The **probability of a favorable outcome** is defined as follows:

$$\text{probability of a favorable outcome} = \frac{\text{number of favorable outcomes}}{\text{total number of possible outcomes}}$$

The **number of favorable outcomes** is simply the number of ways that the favorable outcome can occur.

Expressing Probability as a Number

Probabilities are expressed as numbers (usually common fractions) ranging from 0 to 1 or as percents from 0% to 100%.

Example 1

What is the probability that the spinner pictured above will stop on green?

Step 1. Notice there are 4 *possible outcomes*, but only 1 *favorable outcome*—only 1 green section.

Step 2. Write the probability fraction: $\frac{\text{favorable outcomes}}{\text{possible outcomes}} = \frac{1}{4}$ (one in four)

Answer: The probability of a green outcome is $\frac{1}{4}$, or 25%.
On the average, only **1 spin in 4** will stop on green.

A *probability of 0 (0%)* means that an event cannot occur. The probability is 0 that the spinner will stop on brown because there is no brown section!

A *probability of 1 (100%)* means that an outcome will definitely occur. The probability is 1 that the spinner will stop somewhere in the circle. There is no other possibility, assuming the spinner can't keep spinning forever!

All probabilities you'll ever work with will be between 0 and 1.
- A probability smaller than $\frac{1}{2}$ means that an event will happen less than half of the time—the smaller the probability, the less likely that the event will happen.
- A probability larger than $\frac{1}{2}$ means that an event will happen more than half of the time—the larger the probability, the more likely that the event will happen.

Example 2

Look at the spinner at right. What is the probability that this spinner will stop on a section labeled $10?

Step 1. Notice there are *6 possible outcomes*. Of these 6 outcomes, 2 are labeled $10. Thus, the number of *favorable outcomes* is 2.

Step 2. Write the probability fraction. Reduce this fraction.

$$\frac{\text{favorable outcomes}}{\text{possible outcomes}} = \frac{2}{6} = \frac{1}{3} \text{ (one in three)}$$

On the average, 1 spin in 3 will stop on a $10 section.

Answer: The probability of a $10 outcome is $\frac{1}{3}$, or $33\frac{1}{3}$ %.

Practice

Solve each problem below.

1. There are 6 faces on a die (one of a pair of dice). Each side is equally likely to be "up" after the die is tossed.

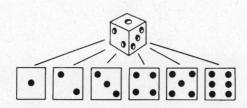

   a) What is the probability of rolling a 4 with one toss of the die?
   b) What is the probability of rolling an even number (2, 4, or 6) with one toss of the die?

2. Below is a spinner for a money game.

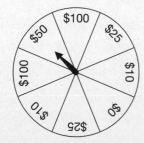

   a) What is the probability that a player will win $25 on one spin?
   b) What is the probability that a player will win $50 or more on one spin?

3. A box contains 4 pennies, 12 nickels, and 20 dimes. Suppose you reach into the box and randomly take out 1 coin. What is the probability that the coin you choose will be

   a) a penny?
   b) a nickel?
   c) a dime?
   d) a quarter?

4. If you randomly pick a number from 1 to 30, what is the probability that the number you pick will be evenly divisible by 4?

5. Suppose you shut your eyes and choose a penny from the group below.

   a) What is the probability that you will choose the "heads up" penny?
   b) What is the probability that you will *not* choose the "heads up" penny?

Problem Solver: **Making a List to Predict Outcomes**

For many problems, making a list of possible outcomes is the best first step. Writing information in this organized way makes it much easier for you to count the number of favorable outcomes.

Example

When you roll a pair of dice, what is the probability that the sum of the two dice will be 7? Asked another way, "What is the probability that you will roll a 7?"

In this problem, each outcome consists of the sum of the two numbers shown face up on the rolled dice. We will first write a list of all combinations of pairs of numbers that the dice can have.

Step 1. Make a list of all possible outcomes. Notice that for each number that die #1 can have, there are 6 values for die #2. As you see, there are 36 possible outcomes.

Die #1	Die #2	Die #1	Die #2	Die #1	Die #2
1	1	3	1	5	1
1	2	3	2	5	2
1	3	3	3	5	3
1	4	3	4	5	4
1	5	3	5	5	5
1	6	3	6	5	6
2	1	4	1	6	1
2	2	4	2	6	2
2	3	4	3	6	3
2	4	4	4	6	4
2	5	4	5	6	5
2	6	4	6	6	6

Step 2. Count the number of favorable outcomes—the number of ways a sum of 7 can occur. In the list above, a box is drawn around each of these. As shown, there are 6 favorable outcomes.

Step 3. Write the probability fraction:

$$\frac{\text{favorable outcomes}}{\text{total outcomes}} = \frac{6}{36} = \frac{1}{6}$$

Answer: The probability of rolling a 7 is $\frac{1}{6}$.

For comparison, notice that there is only 1 way to roll a 2—each die coming up 1. Thus, you are 6 times as likely to roll a 7 as you are to roll a 2.

Practice

■ **Solve each problem below.**

1. Look at the list of outcomes in the example on the previous page.

   a) When you roll a pair of dice, what is the probability that you will roll an 8?

   b) When you roll a pair of dice, what is the probability that you will roll an 11?

2. Amy, Gail, and Vicki are going to a play. They have reserved seats #26, #27, and #28 in the 4th row.

   a) Complete the list below to determine how many different ways the girls can be seated.

#26	#27	#28
Amy	Gail	Vicki
Amy	Vicki	Gail
Gail	etc.	

   b) If each girl randomly chooses a ticket, what is the probability that Amy will sit next to Vicki?

3. At his restaurant, Julian serves orange, cola, and root beer soft drinks. He also serves hamburgers, fish sandwiches, and chicken sandwiches.

   a) How many soft drink–sandwich combinations are possible at Julian's?

   b) If each combination is equally likely to be ordered, what is the probability that Julian's next customer will order a root beer and a fish sandwich?

4. Look again at the list of outcomes in the example on the previous page.

   a) When you roll a pair of dice, how many more times likely are you to roll a 7 than a 4?

   b) When you roll a pair of dice, what is the probability you will roll a "pair" (two identical numbers)?

5. Phil is going to place three posters side by side across the store window. The posters will advertise new styles in ties, belts, and shirts.

   a) Write a list below to show the number of different ways Phil can arrange these posters.

1st	2nd	3rd

   b) If Phil randomly chooses how to arrange the posters, what's the probability that the tie poster will *not* be next to the shirt poster?

6. The first three letters of a license plate were reported to be G, N, and B, but the caller couldn't remember the exact order.

   a) How many different orderings of these 3 letters are possible?

   b) What is the probability that the correct number starts with N?

   c) What is the probability that the correct number starts with either a G *or* a B?

Finding the Probability of Successive Events

To find the probability of **successive events** (events happening one after another), multiply the probability of the first event by the probability of the second event.

Example 1

When rolling a die, what is the probability of rolling two 6s in a row?

On the average, rolling

followed by

happens only once every 36 times you roll the die twice in a row.

Step 1. Realize that the probability of rolling a 6 on either roll is $\frac{1}{6}$.

Step 2. Multiply the first probability ($\frac{1}{6}$) by the second ($\frac{1}{6}$):
$$\frac{1}{6} \times \frac{1}{6} = \frac{1}{36}$$

Answer: The probability of rolling two 6s is $\frac{1}{36}$.

Example 2

If you randomly choose 2 cards from the 5 cards shown below, what is the probability that you will choose 2 face cards?

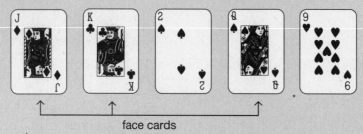

face cards

Step 1. Notice that for your first choice there are 5 possible outcomes, 3 of which are face cards. So, the probability of drawing a face card on your first draw is $\frac{3}{5}$.

Step 2. Now assume you do draw a face card on your first draw. This means that there are 4 possible outcomes for your second draw, only 2 of which are face cards. The probability of drawing a face card on your second draw is then $\frac{2}{4} = \frac{1}{2}$.

Step 3. Multiply the first probability ($\frac{3}{5}$) by the second ($\frac{1}{2}$): $\frac{3}{5} \times \frac{1}{2} = \frac{3}{10}$.

Answer: The probability of choosing 2 face cards is $\frac{3}{10}$.
On the average, you'll draw 2 face cards **3 out of every 10 times** you draw two cards in a row.

Practice

Solve each problem below. The key to success with these problems is to carefully determine the probability of both the first and second events before multiplying.

1. When you are rolling a die, what is the probability of rolling a 5 and then rolling the die again and getting a 2?

2. The median high temperature in June in Salem is equal to 82°F. Estimate the probability that on two days in a row in June the temperature will rise above 82°F. Assume that the daily high temperature is equally likely to be higher or lower than 82°F.

3. A coin has two sides: heads and tails.

   a) If you flip a coin twice in a row, what is the probability of getting 2 heads?
   b) If you flip a coin three times in a row, what is the probability of getting 3 heads?

Problem 4 refers to the drawing below.

4. Suppose you randomly choose 2 cards from the group of cards shown above.

   a) What is the probability that the first card you choose will be a face card?
   b) What is the probability that both cards will be face cards?

5. For Christmas, 7 friends agree to exchange gifts. Each person writes his or her name on a slip of paper. Each then randomly chooses the name of a friend. There are 4 women and 3 men in the group. What is the probability that the first two people who draw will draw a man's name?

6. Audrey's young son made up a game. He has a bag of marbles containing 8 blue marbles and 4 green marbles. He tells Audrey to close her eyes and then choose 2 marbles from the bag. What is the probability that Audrey can win the game by drawing one marble of each color? (Hint: You can assume that Audrey gets *either* a blue marble *or* a green marble on her first draw.)

7. In Bill's pocket are 5 dimes, 3 nickels, and 2 pennies. If Bill randomly takes 2 coins out of his pocket, what is the probability that he will take out 15¢?

Problem 8 refers to the following picture.

8. With your eyes closed, suppose you choose 3 pennies from the group above.

   a) What is the probability that the first 2 pennies you choose will be heads up?
   b) What is the probability that all 3 pennies will be heads up?

Problem Solver: Using Probability for Prediction

Probability is often used as a basis for making predictions. Often, you make many tries (such as rolling a die several times in a row) to see how many times an event (such as rolling a 6) actually occurs. To predict the number of times the event will occur, you multiply the probability of the single event by the number of tries.

Example 1

If you roll a die 300 times, how many 6s will you most likely roll?

Step 1. Determine the probability of rolling a 6 on one roll:

$$\frac{\text{favorable outcomes}}{\text{total outcomes}} = \frac{1}{6}$$

Step 2. Multiply 300 by $\frac{1}{6}$.

$$300 \times \frac{1}{6} = 50$$

Answer: You will most likely roll **50 6s.**

Note: If each of 100 people rolls a die 300 times, not one may roll exactly 50 6s. But the average would likely be 50.

Example 2

If you roll a pair of dice 300 times, how many double 6s will you most likely roll?

Step 1. From the example on page 166, the probability of rolling 2 6s on 1 roll is $\frac{1}{36}$.

Step 2. Multiply 300 by $\frac{1}{36}$.

$$300 \times \frac{1}{36} = 8\frac{1}{3}$$

Answer: You will most likely roll **8 double 6s.**

Note: Example 2 requires a whole-number answer. In probability problems such as this one, round your result to the nearest whole number.

Basing Probability on an Outcome Pattern

Up to now, you've studied how probability can be based on the laws of chance. Now, we'll show how probability can also be based on an **outcome pattern**—a pattern formed by a large number of previous outcomes. The pattern gives the probability of a similar future outcome.

Example 3

Clifford, a basketball player, has made 70% of his free throws so far this season. What is the probability that Clifford will make his first two free throws tonight?

Step 1. Due to Clifford's past performance, we say that the probability he will hit any 1 free throw is 70% (or $\frac{7}{10}$).

Step 2. The probability that Clifford will make 2 in a row is found by multiplying probabilities:

$$\frac{7}{10} \times \frac{7}{10} = \frac{49}{100} \leftarrow \text{probability of making both free throws}$$

probability of making 1st ⬏   ⬑ probability of making 2nd

Answer: $\frac{49}{100}$ **or about 50%** (About half the time, Clifford will make 2 in a row!)

Practice

Solve each problem.

1. If you roll a die 500 times, how many 3s will you most likely roll?

2. On a certain production line, 9 out of the last 200 microchips were defective.

  a) What is the probability that the next microchip will be defective?

  b) Out of the next 750 microchips, what is the most likely number that will be defective?

3. A survey of 1,000 people had these results: 180 preferred bacon and eggs for breakfast, 250 preferred pancakes, 380 preferred cereal, and 190 had other choices.

  a) Based on this survey, what is the probability that the next person asked will prefer pancakes for breakfast?

  b) Of the next 200 people surveyed, how many most likely will say they prefer cereal for breakfast?

Problem 4 refers to the following graph.

Garner Motors Sales

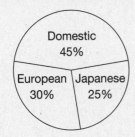

4. a) What is the probability that the next car Garner Motors sells will be a foreign car (not domestic)?

  b) If Garner Motors sells 75 cars next month, how many most likely will be foreign cars?

5. If you roll a pair of dice 250 times, how many double 1s will you most likely roll?

6. Georgia, a softball pitcher, has struck out 10% of the batters she's faced this season.

  a) What is the probability that Georgia will strike out the next batter she pitches to?

  b) What is the probability that Georgia will strike out the next two batters she pitches to?

7. The weather bureau reports that there is a 40% chance for rain tomorrow in Bend and a 25% chance for rain in Newport. The same forecast has already been made 30 times this year!

  a) If the forecast is correct, what is the probability that it will rain in both Bend *and* Newport tomorrow?

  b) How many times this year did it most likely rain in both cities the day after this forecast?

Problem 8 refers to the graph below.

Project Completion Report of WIX Co.

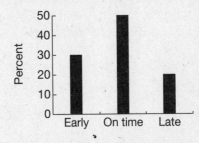

8. a) Of the next 20 projects undertaken by WIX Co., how many most likely will be completed late?

  b) What is the probability that the next 2 projects undertaken by WIX Co. will be completed early (ahead of schedule)?

Test Readiness Checkup

Circle each correct answer. Check your answers on page 248; then correct any errors.

Problems 1–3 refer to the line graph.

Median Family Income in U.S.

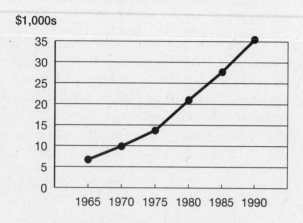

$1,000s

Problems 4–6 refer to the bar graph.

U.S. Households with Selected Media

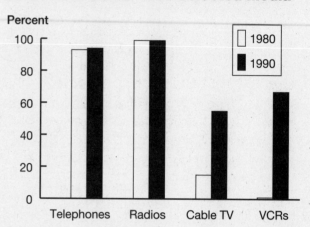

Percent

1. What was the approximate median family income in the U.S. in 1975?

   (1) $10,000
   (2) $15,000
   (3) $20,000
   (4) $25,000
   (5) $30,000

2. What was the approximate median family income in the U.S. in 1983?

   (1) $15,000
   (2) $20,000
   (3) $25,000
   (4) $30,000
   (5) $35,000

3. If the present trend continues, what will be the approximate median family income in the U.S. in the year 2005?

   (1) $30,000
   (2) $40,000
   (3) $50,000
   (4) $60,000
   (5) $70,000

4. Which conclusion is best supported by data on the bar graph above?

   (1) Radio was invented before TV.
   (2) The cost of cable TV is on the rise.
   (3) More families have a radio than a TV.
   (4) The cost of VCRs is decreasing.
   (5) VCRs are more popular than cable TV.

5. Which conclusion is *not* supported by data on the bar graph above?

   (1) The cost of watching movies on a VCR is less than watching them on cable TV.
   (2) Between 1980 and 1990, the percent of American households with a radio changed little, if any.
   (3) By 1990, cable TV was in more than half of all American households.
   (4) In 1990, more American households had a radio than a telephone.
   (5) In 1990, more American households had a VCR than cable TV.

6. You can reasonably predict that by the year 2000, the percent of American households with VCRs will be

   (1) about 10%
   (2) about 30%
   (3) about 50%
   (4) about 70%
   (5) 90% or more

170

7. Joey bought a bag of Popsicles. Three are orange, 5 are grape, and the remaining 4 are cherry. In the car, Joey's daughter reaches into the bag and takes a Popsicle without looking. What is the probability that the Popsicle she takes will be cherry?

(1) $\frac{1}{5}$ (4) $\frac{4}{9}$

(2) $\frac{1}{4}$ (5) $\frac{1}{2}$

(3) $\frac{1}{3}$

8. Tim has 3 shirts: 1 blue, 1 white, and 1 brown. He also has 3 ties: 1 striped, 1 plain, and 1 flowered. If Tim randomly chooses a shirt and tie, what is the probability that he'll wear the white shirt with the striped tie?

(1) $\frac{1}{9}$ (4) $\frac{2}{5}$

(2) $\frac{1}{6}$ (5) $\frac{3}{8}$

(3) $\frac{1}{4}$

10. Following a TV call-in show, 240 viewers phoned the station. Of these, 160 prefer a 1-hour news show, 60 prefer a half-hour news show, and 20 have no preference. Of the next 100 callers, how many most likely will prefer the half-hour news show?

(1) 15   (4) 60

(2) 25   (5) 90

(3) 40

11. On the first day of registration, 25 boys and 25 girls signed up for swimming lessons. Approximately what is the probability that the next two people who register will both be girls?

(1) 10%   (4) 75%

(2) 25%   (5) 90%

(3) 50%

Problem 9 is based on the following picture.

9. Suppose you randomly choose 2 quarters from the group above. What is the probability that both quarters will be heads up?

(1) 10%   (4) 30%

(2) 20%   (5) 50%

(3) 25%

Problem 12 is based on the circle graph.

Age of Students at City College

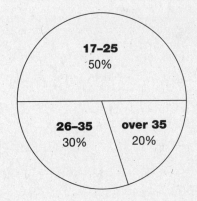

12. What is the probability that the next two students who register at City College will be over 35 years of age?

(1) $\frac{1}{25}$ (4) $\frac{2}{5}$

(2) $\frac{1}{10}$ (5) $\frac{7}{10}$

(3) $\frac{1}{5}$

UNIT 8 Algebra

Algebra is a powerful problem-solving tool that is the basic mathematical language of all technical fields, from auto mechanics to nursing to electronics. In algebra, you use a letter, called a **variable,** to stand for each unknown quantity you're trying to find.

Basic Algebra Equations

Here are examples of the four basic algebra equations written with variables.

Basic Algebra Equations	Solutions	Check
Addition equation: $x + 4 = 9$	$x = 5$	$5 + 4 = 9$
Subtraction equation: $y - 3 = 12$	$y = 15$	$15 - 3 = 12$
Multiplication equation: $2z = 14$	$z = 7$	$2(7) = 14$
Division equation: $\frac{n}{7} = 5$	$n = 35$	$\frac{35}{7} = 5$

Finding the value of the variable that makes the equation a true statement is called **solving the equation.** The correct value of the variable is called the **solution.**

To *check a solution,* substitute the solution for the unknown in the original equation. If the equation is a true statement, then your solution is correct.

Solving Addition and Subtraction Equations

To solve an addition equation, subtract the added number from *each* side of the equation so that the variable is alone and its value found.

Example 1. Solve: $x + 8 = 14$
Step 1. Subtract 8 from each side of the equation.
Step 2. Simplify each side. $(8 - 8 = 0)$
Answer: $x = 6$

$$x + 8 = 14$$
$$x + 8 - 8 = 14 - 8$$
$$x = 6$$

> To simplify an equation, drop expressions such as $8 - 8$ that equal 0.

Check the answer by substituting 6 back into the original equation.

$$6 + 8 = 14$$
$$✔ 14 = 14$$

To solve a subtraction equation, add the subtracted number to *each* side of the equation.

Example 2. Solve: $y - 7 = 9$
Step 1. Add 7 to each side of the equation.
Step 2. Simplify each side. $(-7 + 7 = 0)$
Answer: $y = 16$

$$y - 7 = 9$$
$$y - 7 + 7 = 9 + 7$$
$$y = 16$$

Check: $16 - 7 = 9$
$$✔ 9 = 9$$

Practice

■ **Check to see if the value given for the variable is the solution to each equation. Circle Yes if the given value is the solution and No if it is not.**

1. $x + 7 = 9$ Try $x = 3$ Yes   No

2. $y - 8 = 24$ Try $y = 16$ Yes   No

3. $n + 5 = 13$ Try $n = 8$ Yes   No

4. $a - 4 = 8$ Try $a = 12$ Yes   No

■ **Solve each addition equation by *subtracting the added number from each side of the equation*. Check each answer. The first problem in each row is partially completed.**

5. $x + 4 = 12$ $y + 5 = 9$ $n + 2 = 5$ $m + 8 = 16$
    $x + 4 - 4 = 12 - 4$

6. $y + 9 = 15$ $z + 7 = 19$ $x + 12 = 30$ $n + 3 = 13$
    $y + 9 - 9 = 15 - 9$

7. $z + \frac{3}{2} = 4$ $x + \frac{2}{3} = 2$ $y + \frac{7}{5} = \frac{11}{5}$ $n + 1\frac{3}{4} = 3\frac{2}{3}$
    $z + \frac{3}{2} - \frac{3}{2} = 4 - \frac{3}{2}$

■ **Solve each subtraction equation by *adding the subtracted number from each side of the equation*. Check each answer.**

8. $x - 8 = 7$ $y - 5 = 12$ $z - 3 = 7$ $n - 2 = 0$
    $x - 8 + 8 = 7 + 8$

9. $m - 7 = 13$ $x - 1 = 17$ $y - 14 = 3$ $p - 4 = 4$
    $m - 7 + 7 = 13 + 7$

10. $x - \frac{5}{8} = 3$ $y - \frac{1}{3} = 2$ $z - \frac{5}{4} = \frac{15}{8}$ $n - 2\frac{3}{4} = 3\frac{1}{8}$
    $x - \frac{5}{8} + \frac{5}{8} = 3 + \frac{5}{8}$

Solving Multiplication and Division Equations

To solve a multiplication equation, divide each side of the equation by the number that multiplies the variable.

Example 1. Solve: $6x = 42$

$$6x = 42$$

Step 1. Divide each side of the equation by 6.

Step 2. Simplify each side.

$$\frac{6x}{6} = \frac{42}{6}$$

$$x = 7$$

(The 6s on the left cancel out because $\frac{6}{6} = 1$. 1x is the same as x.)

Answer: $x = 7$

Check the answer by substituting 7 for the unknown (x) in the original equation.

$$6(7) = 42$$
$$✔42 = 42$$

To solve a division equation, multiply each side of the equation by the number that divides the variable.

Example 2. Solve $\frac{n}{4} = 9$

$$\frac{n}{4} = 9$$

Step 1. Multiply each side of the equation by 4.

Step 2. Simplify each side.

$$\frac{n}{4}(4) = 9(4)$$

$$n = 36$$

(The 4s on the left cancel out because $\frac{1}{4} \times 4 = 1$.)

Answer: $n = 36$

$$\text{Check: } \frac{36}{4} = 9$$
$$✔9 = 9$$

If a fraction multiplies an unknown, multiply each side of the equation by the **reciprocal** of the fraction. The reciprocal of a fraction is the new fraction you get by switching the numerator and denominator.

Example 3. Solve: $\frac{2}{3}x = 10$

$$\frac{2}{3}x = 10$$

Step 1. Multiply each side of the equation by the reciprocal of $\frac{2}{3}$, which is $\frac{3}{2}$.

Step 2. Simplify each side.

$$\frac{3}{2}\left(\frac{2}{3}\right)x = \frac{3}{2}(10)$$

$$x = 15$$

(On the left, $\frac{3}{2}$ cancels $\frac{2}{3}$ because $\frac{3}{2} \times \frac{2}{3} = 1$.)

Answer: $x = 15$

$$\text{Check: } \frac{2}{3}(15) = 10$$
$$✔10 = 10$$

Practice

▋ **Check to see if the value given for the variable is the solution to each equation. Circle Yes if the given value is the solution and No if it is not.**

1. $7x = 63$ Try $x = 8$ Yes   No

2. $\frac{y}{4} = 12$ Try $y = 48$ Yes   No

3. $\frac{2}{3}z = 12$ Try $z = 21$ Yes   No

▋ **Solve each multiplication equation by *dividing each side of the equation by the number that multiplies the variable*. Check each answer. The first problem in each row is partially completed.**

4. $2y = 14$ $3z = 24$ $5n = 45$ $7z = 42$

$\dfrac{2y}{2} = \dfrac{14}{2}$

5. $2x = 3.5$ $4y = 10.4$ $3z = \frac{13}{2}$ $5n = 3\frac{1}{2}$

$\dfrac{2x}{2} = \dfrac{3.5}{2}$

▋ **Solve each division equation by *multiplying each side of the equation by the number that divides the variable*. Check each answer.**

6. $\frac{x}{3} = 12$ $\frac{y}{5} = 9$ $\frac{z}{7} = 5$ $\frac{n}{6} = 14$

$\dfrac{x}{3}(3) = 12(3)$

7. $\frac{n}{2} = 2\frac{1}{4}$ $\frac{x}{3} = 4.5$ $\frac{y}{4} = 3\frac{3}{5}$ $\frac{p}{7} = 2.8$

▋ **Solve each equation by *multiplying each side of the equation by the reciprocal of the fraction that multiplies the variable*. Check each answer.**

8. $\frac{2}{5}x = 14$ $\frac{3}{4}y = 12$ $\frac{7}{8}z = 14$ $\frac{11}{16}n = 33$

$\dfrac{5}{2}\left(\dfrac{2}{5}\right)x = \dfrac{5}{2}(14)$

Solving Multi-Step Equations

A **multi-step equation** contains two or more of the operations of addition, subtraction, multiplication, and division.

To solve a multi-step equation *containing variables*, follow these two rules:

1. Do addition or subtraction first.

2. Do multiplication or division last.

Example 1. Solve: $3y - 9 = 36$

$$3y - 9 = 36$$

Step 1. Add 9 to each side of the equation.
    Simplify each side.

$$3y - 9 + 9 = 36 + 9$$
$$3y = 45$$

Step 2. Divide each side of the equation by 3.
    Simplify each side.

$$\frac{3y}{3} = \frac{45}{3}$$
$$y = 15$$

Answer: $y = 15$

Check the answer by substituting 15 for the unknown in the original equation.

$$3(15) - 9 = 36$$
$$45 - 9 = 36$$
$$✔36 = 36$$

Example 2. Solve: $\frac{x}{5} + 6 = 8$

$$\frac{x}{5} + 6 = 8$$

Step 1. Subtract 6 from each side of the equation.
    Simplify each side.

$$\frac{x}{5} + 6 - 6 = 8 - 6$$
$$\frac{x}{5} = 2$$

Step 2. Multiply each side of the equation by 5.
    Simplify each side.

$$\frac{x}{5}(5) = 2(5)$$
$$x = 10$$

Answer: $x = 10$

Check:
$$\frac{10}{5} + 6 = 8$$
$$2 + 6 = 8$$
$$✔8 = 8$$

Practice

■ **Solve each equation. Remember: At each step, you must perform the same operation on each side of the equation.**

1. $2x + 8 = 10$ $3y - 4 = 8$ $4n + 7 = 11$ $5z - 9 = 36$

2. $8m + 11 = 43$ $4p - 9 = 27$ $7y + 3.5 = 17.5$ $5x = 12\frac{1}{2}$

3. $\frac{x}{5} + 4 = 7$ $\frac{y}{3} - 7 = 3$ $\frac{z}{6} + 8 = 9$ $\frac{n}{4} - 8 = 12$

4. $\frac{z}{3} - 1 = 6.5$ $\frac{m}{4} + \frac{3}{2} = \frac{8}{2}$ $\frac{y}{5} - 3\frac{1}{3} = 2\frac{2}{3}$ $\frac{p}{6} + 8 = 14\frac{1}{4}$

5. $\frac{3}{4}x + 7 = 16$ $\frac{2}{3}y - 9 = 9$ $\frac{4}{5}z + 12 = 20$ $\frac{8}{5}n - 4 = 20$

6. $\frac{7}{4}x - 9 = 26$ $\frac{4}{3}z + \frac{5}{3} = \frac{14}{3}$ $2\frac{1}{2}x + 4 = 19$ $4\frac{2}{3}y - 7 = 21$

(Hint: Write $2\frac{1}{2}$ as $\frac{5}{2}$.)

Applying Your Skills

Learning to use equations to solve word problems is a very important part of your study of algebra. On these next two pages, you'll learn to write and solve equations for problems that you've already learned to solve—without using algebra! See if algebra simplifies these problems.

To solve a word problem using algebra, follow these steps:
Step 1. Represent the unknown with a letter (usually the letter x).
Step 2. Write an equation that represents the given information.
Step 3. Solve the equation to find the unknown value.

Example 1. Five times a number is equal to 75. What is the number?

Step 1. Let x = unknown number
Step 2. Write an equation for x: $5x = 75$
Step 3. Solve the equation.
  Divide each side by 5; then simplify.

Solve:
$$5x = 75$$
$$\frac{5x}{5} = \frac{75}{5}$$
$$x = 15$$

Answer: $x = 15$

The unknown number is 15.

Check: $5(15) = 75$
✔ $75 = 75$

Example 2. Muriel earns \$36 each day in salary as a waitress. She also earns \$2.50 per hour in tips. On Tuesday, Muriel earned a total of \$51. How many hours did Muriel work on Tuesday?

Step 1. Let x = number of hours Muriel worked on Tuesday.
Step 2. Write an equation for x: $\$2.50x + \$36 = \$51$
Step 3. Solve the equation.
  a) Subtract \$36 from each side; then simplify.

  b) Divide each side by \$2.50; then simplify.

Solve:
$$\$2.50x + \$36 = \$51$$
$$\$2.50x + \$36 - \$36 = \$51 - \$36$$
$$\$2.50x = \$15$$
$$\frac{\$2.50x}{\$2.50} = \frac{\$15}{\$2.50}$$
$$x = 6$$

Answer: $x = 6$ hours

Check:
$$\$2.50(6) + \$36 = \$51$$
$$\$15 + \$36 = \$51$$
✔ $\$51 = \$51$

Muriel worked for 6 hours on Tuesday.

Practice

▨ In each problem below, do the following:

a) Assign the unknown value a variable.
b) Write an equation using the variable.
c) Solve the equation by finding the value of the unknown.

1. One-fourth of Kari's salary goes for rent. If Kari's monthly rent payment is $375, what is Kari's monthly salary?

   a) Variable: x = Kari's monthly salary

   b) Equation:

   c) Solution:

2. Joseph paid $320 for a new color TV. If Joseph was given a $65 discount, what was the original price of the set?

   a) Variable: p = original price of set

   b) Equation:

   c) Solution:

3. For working 8 hours, Vince was paid $71. Of this amount, $15 was a bonus. What was Vince's regular hourly rate?

   a) Variable: r = regular hourly rate

   b) Equation:

   c) Solution

4. Each month, Wendy deposits $\frac{1}{10}$ of her check in a savings account. After 6 deposits, she placed an additional $150 in this account. If she now has $900 in savings, how much does Wendy usually deposit each month?

   a) Variable: d =

   b) Equation:

   c) Solution:

5. At Hoover Elementary School, 60% of the students are in the grades kindergarten through third grade. If 180 children are in these 4 grades, what is the total enrollment at Hoover Elementary?

   a) Variable:

   b) Equation:

   c) Solution:

6. Jackie cut a ribbon into 8 equal pieces. She measured and found that each piece was two inches longer than a foot-long ruler. What was the length of the original uncut ribbon?

   a) Variable:

   b) Equation:

   c) Solution:

7. Three-fourths of the punch was gone during the first hour of the barbecue. To the remaining punch, Jamie added 3 quarts of newly made punch. If the total amount of punch is now 7 quarts, how much punch was there at the start of the barbecue?

   a) Variable:

   b) Equation:

   c) Solution:

Combining Like Terms in an Equation

An equation is made up of **terms**. Each term is a number standing alone or is a variable (representing an unknown value) multiplied by a **coefficient** (a number that is constant no matter what value the variable is assigned).

When the same variable appears in more than one term in an addition or subtraction equation, you combine the separate *like terms* by adding or subtracting the coefficients. The following examples show how separate variables are combined:

$4x + 3x = 7x$ since $4 + 3 = 7$

$8y - 2y = 6y$ since $8 - 2 = 6$

$\frac{3}{2}n + n = \frac{5}{2}n$ since $\frac{3}{2} + 1 = \frac{3}{2} + \frac{2}{2} = \frac{5}{2}$

$9z - z = 8z$ since $9 - 1 = 8$

Note: When a variable stands alone, the coefficient is understood to be 1.
Therefore, in the examples shown above, the coefficient of n is 1 and the coefficient of z is 1.

Example 1. Solve: $7x + 3x = 100$ $7x + 3x = 100$

Step 1. Combine like terms. $10x = 100$

Step 2. Divide each side by 10. $\frac{10x}{10} = \frac{100}{10}$
　　　Simplify each side. $x = 10$

Answer: $x = 10$

$$\text{Check: } 7(10) + 3(10) = 100$$
$$70 + 30 = 100$$
$$✔\, 100 = 100$$

Example 2. Solve: $12n - 4n + 10 = 42$ $12n - 4n + 10 = 42$

Step 1. Combine like terms. $8n + 10 = 42$

Step 2. Subtract 10 from each side. $8n + 10 - 10 = 42 - 10$
　　　Simplify each side. $8n = 32$

Step 3. Divide each side by 8. $\frac{8n}{8} = \frac{32}{8}$
　　　Simplify each side. $n = 4$

Answer: $n = 4$

$$\text{Check: } 12(4) - 4(4) + 10 = 42$$
$$48 - 16 + 10 = 42$$
$$32 + 10 = 42$$
$$✔\, 42 = 42$$

Practice

◼ Combine each pair of like terms below.

1. $3x + 2x$ $7n - 4n$ $9y + y$ $12z - 2z$

2. $\frac{1}{2}x + \frac{1}{4}x$ $\frac{11}{8}y - y$ $\frac{2}{3}n + \frac{1}{2}n$ $1\frac{5}{8}w - \frac{6}{16}w$

◼ Solve each equation below.

3. $6x + 2x = 32$ $5y - y = 12$ $4z + 9z = 39$ $9m - 4m = 45$

4. $\frac{2}{3}x + \frac{2}{3}x = 3$ $\frac{7}{8}n - \frac{3}{8}n = 2$ $\frac{7}{5}z - \frac{3}{10}z = 4$ $\frac{3}{4}x - \frac{2}{3}x = 1$

5. $1.4z + 0.6z = 17$ $2.6x - 0.7x = 38$ $6y - 3.75y = 9$ $4.5x + 3x = 61.5$

6. $9n - 4n + 18 = 28$ $2z - z + 4 = 11$ $5y + 2y - 5 = 7$ $12x + 3x + 14 = 76$

7. $3x + x - 9 = 47$ $8q - 3q + 15 = 40$ $m + 3m + 6 = 18$ $5n + 3n - 12 = 8$

Solving Equations with Terms on Both Sides

A variable (representing an unknown value) can appear in terms on both sides of an equation. Usually, you'll move all terms containing a variable to the left side of the equation.

Example 1. Solve: $5x = 3x + 12$

$$5x = 3x + 12$$

Step 1. Subtract $3x$ from each side.
Simplify each side.

$$5x - 3x = 3x - 3x + 12$$
$$2x = 12$$

Step 2. Divide each side by 2.
Simplify each side.

$$\frac{2x}{2} = \frac{12}{2}$$
$$x = 6$$

Answer: $x = 6$

Check: $5(6) = 3(6) + 12$
$$30 = 18 + 12$$
$$\checkmark 30 = 30$$

Example 2. Solve: $6y - 14 = 2y + 82$

$$6y - 14 = 2y + 82$$

Step 1. Subtract $2y$ from each side.
Simplify each side.

$$6y - 2y - 14 = 2y - 2y + 82$$
$$4y - 14 = 82$$

Step 2. Add 14 to each side.
Simplify each side.

$$4y - 14 + 14 = 82 + 14$$
$$4y = 96$$

Step 3. Divide each side by 4.
Simplify each side.

$$\frac{4y}{4} = \frac{96}{4}$$
$$y = 24$$

Answer: $y = 24$

Check: $6(24) - 14 = 2(24) + 82$
$$144 - 14 = 48 + 82$$
$$\checkmark 130 = 130$$

Practice

▩ Complete each problem below. Check each answer.

1.

$$6x + 9 = 2x + 25$$
$$6x - 2x + 9 = 2x - 2x + 25$$
$$4x + 9 = 25$$
$$4x + 9 - 9 = 25 - 9$$

$$6z - 19 = 3z + 47$$
$$6z - 3z - 19 = 3z - 3z + 47$$
$$3z - 19 = 47$$
$$3z - 19 + 19 = 47 + 19$$

$$11n - 13 = 4n + 36$$
$$11n - 4n - 13 = 4n - 4n + 36$$
$$7n - 13 = 36$$
$$7n - 13 + 13 = 36 + 13$$

Check:

$$6(\ ) + 9 = 2(\ ) + 25$$

Check:

$$6(\ ) - 19 = 3(\ ) + 47$$

Check:

$$11(\ ) - 13 = 4(\ ) + 36$$

2.

$$5y - 7 = 3y + 23$$
$$5y - 3y - 7 = 3y - 3y + 23$$
$$2y - 7 = 23$$

$$4x + 12 = x + 39$$
$$4x - x + 12 = x - x + 39$$
$$3x + 12 = 39$$

$$13p + 28 = 5p + 52$$
$$13p - 5p + 28 = 5p - 5p + 52$$
$$8p + 28 = 52$$

Check:

$$5(\quad) - 7 = 3(\quad) + 23$$

Check:

$$4(\quad) + 12 = (\quad) + 39$$

Check:

$$13(\quad) + 28 = 5(\quad) + 52$$

■ **Solve each equation below. Check each answer.**

3. $11n - 6 = 9n$ $12y = 3y + 18$ $4x - 15 = x$

4. $4b + 17 = b + 23$ $2y + 14 = y + 17$ $5x - 7 = 2x + 33$

5. $9x + 32 = 4x + 82$ $8w - 13 = w + 50$ $3z + 14 = 2z + 14$

6. $\frac{3}{5}n - 2 = \frac{1}{5}n + 3$ $x + 8 = \frac{1}{4}x + 32$ $\frac{7}{8}y - 9 = \frac{3}{4}y + 5$

Applying Your Skills

The following examples will help you understand how an equation is written to represent information given in more complicated word problems.

Example 1. Five times Blake's age minus 2 is equal to 2 times his age plus 10. How old is Blake?

Step 1. Let x equal Blake's age.

Step 2. Write an equation that represents the information given in the problem.

$$5x - 2 = 2x + 10$$

Step 3. Solve the equation:

    Solve: $5x - 2 = 2x + 10$

  a) Add 2 to each side.
     Simplify.

$$5x - 2 + 2 = 2x + 10 + 2$$
$$5x = 2x + 12$$

  b) Subtract $2x$ from each side.
     Simplify.

$$5x - 2x = 2x - 2x + 12$$
$$3x = 12$$

  c) Divide each side by 3.
     Simplify.

$$\frac{3x}{3} = \frac{12}{3}$$
$$x = 4$$

Answer: Blake is 4 years old.

Example 2. Martin has money in a savings account. If he adds $200 each month for the next 8 months, he will have a total then that is three times as much as he has now, not counting any interest. How much is Martin's account worth now?

Step 1. Let x = amount in Martin's account now
    $x + \$1,600$ = amount in Martin's account in 8 months ($\$200 \times 8 = \$1,600$)
        $3x$ = three times the amount he has now

Step 2. Write an equation for the problem.

$$x + \$1,600 = 3x$$

Step 3. Solve the equation:

    Solve: $x + \$1,600 = 3x$

  a) Subtract x from each side.
     Simplify.

$$x - x + \$1,600 = 3x - x$$
$$\$1,600 = 2x*$$

  b) Divide each side by 2.
     Simplify.

$$\frac{\$1,600}{2} = \frac{2x}{2}$$
$$\$800 = x$$

        or $x = \$800$

Answer: Martin has $800 in his account now.

*Note: Sometimes, the variable ends up on the right of the equals sign. Don't let this confuse you. Just solve the equation and, as your final step, write the variable on the left side.

Practice

■ **Solve the following problems using equations.**

1. Represent the following sentence as an equation where *n* stands for the unknown number. Then solve for *n*.

   "Four times a number plus 2 is equal to 2 times the same number plus 8."

2. Letting x = unknown amount, write the following information as an equation. Then solve for x.

   "Three-fourths of an amount plus $150 is equal to twice the original amount."

3. Three numbers add up to 264. The second number is twice as large as the first. The third number is three times as large as the first. What are the three numbers?

4. Anna and Jo went to lunch. The total bill is $8.35. Anna's lunch cost $0.65 more than Jo's. How much is each person's share?

5. Sally is thinking of a number. She says, "If you divide the number by 4 and add 11 to the quotient, the number that results is equal to three times the original number." Which equation can you use to find Sally's number (*n*)?

   (1) $\frac{n}{4} = 3n + 11$

   (2) $\frac{n + 11}{4} = 3n$

   (3) $\frac{n + 4}{11} = 3n$

   (4) $\frac{n}{4} + 11 = 3n$

   (5) $4n + 11 = 3n$

6. Jean and Chris run a cleaning service. Jean provides all the supplies and earns $200 more each month than Chris. In July, the two earned a total of $2,400. Of this amount, how much should each person receive?

7. Consecutive integers are whole numbers that follow one another: for example 7, 8, and 9. If the sum of 3 consecutive integers is 39, what are the integers?

   (Hint: Let x = 1st integer; $x + 1$ = 2nd integer; and $x + 2$ = 3rd integer.)

8. Allison saves stamps. If she saves 25 stamps each month for the next year, she'll end up with 50% more stamps than she has now. How many stamps does Allison have now?

9. Lita makes twice as much profit on each blouse she sells as she does on each skirt. When she sells a blouse and skirt together, she makes a total profit of $10.50. How much profit does Lita make on each item?

10. For every dollar that he makes, Dave's wife earns $1.25. Last month their total monthly income was $2,140. Which equation below can be used to determine Dave's income last month (*x*)?

    (1) $x + \frac{4}{3}x = \$2,140$

    (2) $x + x = \frac{\$2,140}{\$1.25}$

    (3) $x + \frac{5}{4}x = \$2,140$

    (4) $x + x + \$1.25 = \$2,140$

    (5) $x + x + \$0.75 = \$2,140$

Learning About Proportion

A proportion is made up of two equal ratios.* For example, if you mix 2 cups of pineapple juice with 3 cups of orange juice, the ratio of pineapple juice to orange juice is $\frac{2}{3}$. You get the same-flavored mixture by adding 6 cups of pineapple juice to 9 cups of orange juice. The two mixtures have the same flavor because the ratios of ingredients are equal: $\frac{2}{3} = \frac{6}{9}$. The ratios form a proportion.

You write a proportion in symbols in either of two ways:

(1) With colons, 2:3 = 6:9

(2) As equal fractions, $\frac{2}{3} = \frac{6}{9}$

You read a proportion as two equal ratios connected by the word *as*.

2:3 = 6:9 is read "2 is to 3 *as* 6 is to 9."

$\frac{2}{3} = \frac{6}{9}$ is read "2 is to 3 *as* 6 is to 9."

In a proportion, the **cross products** are equal. To find the cross products, cross multiply. Multiply each numerator by the opposite denominator.

Cross Multiplication

$$\frac{2}{3} \diagdown\!\!\!\!\diagup \frac{6}{9}$$

Equal Cross Products

$$2(9) = 3(6)$$

$$18 = 18$$

Often, you are asked to find a missing number in a proportion. To solve this type of problem, use a variable to represent the missing number (*x* or another letter), and cross multiply. Then solve the equation.

Example 1	Example 2
Find the missing term: ___:4 = 12:16	Find the missing term: $\frac{7}{3} = \frac{28}{x}$
Step 1. Write the proportion as equal fractions, with the unknown value written as *x*.　　$\frac{x}{4} = \frac{12}{16}$	*Step 1.* Cross multiply.　$7x = 3(28)$ 　　　　　　　　　　　$= 84$
	Step 2. Solve for *x*.　$x = \frac{84}{7} = 12$
Step 2. Cross multiply.　$16x = 4(12)$ 　　($x16 = 16x$)　　　$= 48$	Answer: *x* = 12
Step 3. Solve for *x*.　$x = \frac{48}{16} = 3$	
Answer: *x* = 3	

*For a review of ratios, reread pages 94 and 95.

Practice

■ Cross multiply to see if each pair of fractions forms a proportion. Remember: In a proportion, the cross products are equal. Circle Yes if it is a true proportion. Circle No if no proportion is formed.

1. $\frac{3}{5} = \frac{20}{35}$ Yes   No $\frac{8}{5} = \frac{24}{16}$ Yes   No $\frac{11}{16} = \frac{33}{48}$ Yes   No

2. $\frac{4}{3} = \frac{28}{21}$ Yes   No $\frac{5}{6} = \frac{32}{36}$ Yes   No $\frac{8}{7} = \frac{56}{49}$ Yes   No

■ Write the following proportions as two equal fractions.

3. a) Five is to six as fifteen is to eighteen.   b) Three is to two as nine is to six.

4. $5:15 = 1:3$ $4:3 = 32:24$ $x:9 = 24:29$ $4:5 = 28:y$

■ Find the missing term in each proportion. When cross multiplying, write the product containing the variable to the left of the = sign.

5. $\frac{x}{4} = \frac{6}{8}$ $\frac{6}{15} = \frac{12}{y}$ $\frac{3}{n} = \frac{12}{8}$ $\frac{20}{25} = \frac{x}{5}$

6. $\frac{h}{9} = \frac{18}{27}$ $\frac{6}{x} = \frac{18}{12}$ $\frac{10}{6} = \frac{15}{y}$ $\frac{15}{16} = \frac{x}{64}$

7. $9:12 = 15:x$ $14:10 = y:5$ $10:6 = 5:h$ $3:8 = x:32$

8. ____:28 = 3:7   4:____ = 12:15   3:2 = ____:48   7:16 = 21:____

Problem Solver: Using Proportions to Solve Word Problems

You can use proportions to solve word problems involving comparisons. As in ratio problems, be sure to write the terms of a proportion in the order stated in each problem.

Example 1. A light blue paint is made by mixing 2 parts of blue paint to 5 parts of white paint. How many gallons of blue paint should be mixed with 24 gallons of white paint to make this light blue color?

Step 1. Write a proportion where each ratio is $\frac{\text{amount of blue}}{\text{amount of white}}$.

Let x stand for the unknown number of gallons of blue.

Required ratio $\rightarrow \frac{2}{5} = \frac{x}{24}$ ← Ratio of mixture being made

Step 2. Cross multiply; then solve the resulting equation for x.
For simplicity, write $5x$ to the left of the = sign.

$$5x = 2(24) = 48$$

$$x = \frac{48}{5} = 9.6 \text{ gallons}$$

Answer: 9.6 gallons

You may find that using proportions simplifies your work with multi-step rate problems.

Example 2. On the first day of her trip, Pam drove 480 miles in 8 hours. Driving at the same rate, how far can Pam drive in $5\frac{1}{2}$ hours on the second day?

Instead of first finding the rate, you can use a proportion to solve this problem.

Step 1. Write a proportion where each ratio is $\frac{\text{miles}}{\text{hours}}$.

Let x stand for the unknown mileage of the second day.
(Notice that a number in the proportion itself may contain a fraction.)

First-day results $\rightarrow \frac{480}{8} = \frac{x}{5\frac{1}{2}}$ ← Second-day results

Step 2. Cross multiply; then solve the resulting equation for x.

$$8x = 480(5\tfrac{1}{2}) = 2{,}640$$

$$x = \frac{2{,}640}{8} = 330 \text{ miles}$$

Answer: 330 miles

Practice

■ **Solve the following problems by using proportions. (Notice that you also know how to solve several of these problems without using proportions.)**

1. A tropical punch recipe calls for 3 parts 7-Up to 4 parts Hawaiian Punch. How much 7-Up should be added to 3 quarts of Hawaiian Punch to make the mixture?

2. Working for 8 hours, Brandon earned $59.20. At this pay rate, how much does Brandon earn in a 40-hour workweek?

3. If a 9-ounce piece of steak costs $1.98, what would be the price of a 13-ounce piece of the same type of steak?

4. Gloria drove her new car 255 miles on just 8 gallons of gas. Given similar driving conditions, how far can Gloria expect to drive on 12 gallons?

5. Which number correctly completes this statement: "If 9 defects are found in 270 television sets, you can expect to find ____ defects in 1,200 television sets."

   (1) 20
   (2) 30
   (3) 40
   (4) 50
   (5) 60

6. While driving across Texas, Carol drove 392 miles on Monday in 7 hours. If she drives at the same rate, how far can Carol expect to drive on Tuesday in 4 hours and 30 minutes?

7. The Delta company has 3 women managers for every 4 men managers. If Delta has 28 men managers, how many women managers does it have?

8. 25.4 centimeters equals exactly 10 inches. Knowing this, determine how many centimeters are in 25 inches.

9. Tough Guard epoxy is supposed to be mixed in the ratio of 3 parts hardener to 8 parts base. How many drops of hardener should you mix with 40 drops of base?

10. In a recent survey of 135 voters, 3 out of every 5 supported raising the gasoline tax. Which of the following proportions can be used to find the number of voters who are *not* in favor of raising this tax?

   (1) $\frac{3}{5} = \frac{135}{x}$
   (2) $\frac{3}{7} = \frac{x}{135}$
   (3) $\frac{3}{5} = \frac{x}{135}$
   (4) $\frac{2}{5} = \frac{135}{x}$
   (5) $\frac{2}{5} = \frac{x}{135}$

Solving an Equation with Two Variables

So far in our study of algebra, we have discussed equations that contain a single variable. Now you'll learn how to solve and graph equations that contain two variables.

As an example, the equation $y = 3x + 2$ contains two variables, y and x.

dependent variable ⌐   ⌐ independent variable

Written as above, the variable to the left of the equals sign (y) is called the **dependent variable**, while the variable to the right (x) is called the **independent variable**.

An equation containing two variables has more than one solution. The value of the dependent variable *depends* on the value you choose for the independent variable. In $y = 3x + 2$, each y value depends on a chosen x value.

To solve an equation with two unknowns, follow these two steps:

Step 1. Choose several values for the independent variable.

Step 2. Substitute each chosen value of the independent variable into the original equation; then find the matching value of the dependent variable.

To keep the solutions in order, write them in a **Table of Values**.

Example

Solve for y in the equation $y = 3x + 2$ for $x = 0, 1, 2,$ and 3.

Step 1. Write the chosen x values in the Table of Values.

Step 2. Substitute each x value into the equation $y = 3x + 2$. For each x value, find the matching y value.

Step 3. Fill in the Table of Values for y.

Choose x value.	Solve for y value.	Table of Values	
x value	$y = 3x + 2$	x	y
a) $x = 0$	$y = 3(0) + 2 = 2$	0	2
b) $x = 1$	$y = 3(1) + 2 = 5$	1	5
c) $x = 2$	$y = 3(2) + 2 = 8$	2	8
d) $x = 3$	$y = 3(3) + 2 = 11$	3	11

Practice

In each equation below, identify both the dependent and the independent variable.

1. a) $y = 7x$ b) $c = 3s - 5$ c) $m = \frac{2}{3}n + 6$

    dependent variable: ____   dependent variable: ____   dependent variable: ____

    independent variable: ____   independent variable: ____   independent variable: ____

In each equation below, find the value of the *dependent variable* for the given value of the *independent variable*.

2. $y = 4x$

   If $x = 3$, $y =$ ____.

   $a = 5b + 9$

   If $b = 0$, $a =$ ____.

   $m = 4n - 8$

   If $n = 5$, $m =$ ____.

Complete each Table of Values by solving each equation for the given values of the independent variable.

3. $y = x + 3$ Table of Values

x	y
0	
1	
2	
3	

6. $c = 4n + 1$ Table of Values

n	c
2	
4	
6	
8	

4. $y = 2x - 2$ Table of Values

x	y
1	
2	
3	
4	

7. $r = \frac{1}{2}s + 6$ Table of Values

s	r
0	
3	
6	
9	

5. $p = 3q + 0.5$ Table of Values

q	p
1.5	
2	
2.5	
3	

8. $m = \frac{2}{3}n - 1$ Table of Values

n	m
3	
6	
9	
12	

Becoming Familiar with a Coordinate Graph

The Coordinate Graph

To graph an equation, you need to become familiar with a **coordinate graph**. A coordinate graph is formed by combining a vertical number line, called the **y axis**, with a horizontal number line, called the **x axis**. (To review negative numbers and number lines, reread pages 16 and 17 at this time.)

The point at which the two axes meet is called the **origin** of the graph. The origin has a value of 0 for both axes.

Reading Points on a Coordinate Graph

Every point on a graph has **coordinates**, two numbers that tell its position.
- The *x coordinate* tells how far the point is from the *y* axis.
  Positive *x* indicates the point is to the right of the *y* axis.
  Negative *x* indicates the point is to the left of the *y* axis.
- The *y coordinate* tells how far the point is from the *x* axis.
  Positive *y* indicates the point is above the *x* axis.
  Negative *y* indicates the point is below the *x* axis.
- Coordinates are usually written in parentheses, the *x* coordinate written first, followed by the *y* coordinate: (*x* coordinate, *y* coordinate).

Example: On the graph below, point A = (3,4).

x coordinate ⬑ ⬐ y coordinate

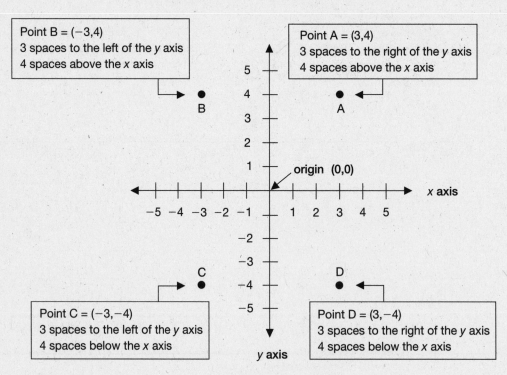

Point B = (−3,4)
3 spaces to the left of the y axis
4 spaces above the x axis

Point A = (3,4)
3 spaces to the right of the y axis
4 spaces above the x axis

origin (0,0)

x axis

Point C = (−3,−4)
3 spaces to the left of the y axis
4 spaces below the x axis

Point D = (3,−4)
3 spaces to the right of the y axis
4 spaces below the x axis

y axis

Coordinate Graph

Plotting Points on a Coordinate Graph

To plot a point on a coordinate graph, follow these steps:

Step 1. Locate the *x* coordinate on the *x* axis.

Step 2. From this point, move directly up (for a positive *y* coordinate) or directly down (for a negative *y* coordinate) to the point that's directly across from the *y* coordinate value.

Example: To plot point A = (4,3), find the coordinate point 4 on the *x* axis. Then move directly up 3 spaces to the point that is across from 3 on the *y* axis. Plot the point; then label it with an A or with its coordinates.

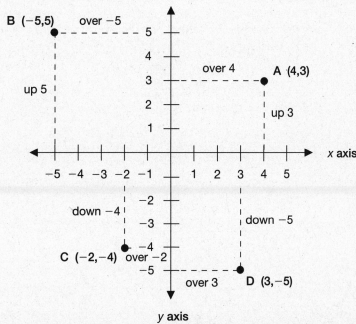

Practice

1. Identify the coordinates of each point plotted below.

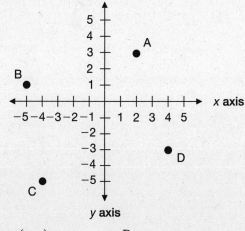

A = (,)   B =

C =   D =

2. Plot each of the points whose coordinates are given below.

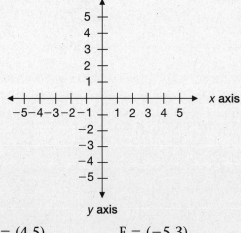

E = (4,5)   F = (−5,3)

G = (−2,−3)   H = (4,−5)

Graphing a Linear Equation

The equation $y = 2x + 1$ is an example of a **linear equation**. When solutions of a linear equation are plotted on a coordinate graph, they always lie on a straight line. Drawing this line of solutions is called *graphing the equation*.

To graph a linear equation, follow these steps:

Step 1. Choose three values for the independent variable. Then determine the matching values for the dependent variable and write each pair of values in a Table of Values.

Step 2. Plot the three points on a coordinate graph; then connect them with a line extending to the edges of the graph.

■ **Math Tip**

Note: Only two points are needed to draw a line, but the third point serves as a check that you plotted the first two points correctly.

Example: Graph the equation $y = 2x + 1$.

Step 1. Let $x = 0$, 1, and 2. Solve the equation $y = 2x + 1$ for these three values. Make a Table of Values.

Table of Values

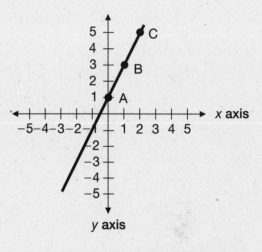

x value	$y = 2x + 1$	x	y
$x = 0$	$y = 2(0) + 1 = 1$	0	1
$x = 1$	$y = 2(1) + 1 = 3$	1	3
$x = 2$	$y = 2(2) + 1 = 5$	2	5

Step 2. Write the pairs of values in the Table of Values as points to be plotted:

A = (0,1)   B = (1,3)   C = (2,5)

Step 3. Plot the points; then connect them with a straight line extending to the edges of the graph.

Practice

■ **Graph each equation on the graph at right. Use the listed values of each independent variable.**

1. $y = 2x - 4$

Table of Values

x	y
2	
3	
4	

2. $y = \frac{1}{2}x + 1$

Table of Values

x	y
0	
2	
4	

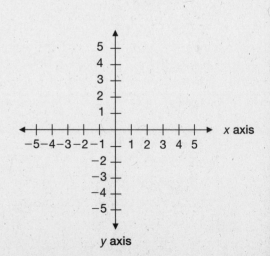

194

3. Blue Cab Company charges a $1.25 pickup charge plus $1.25 per mile for local service. If *f* stands for fare (total cost of the trip) and *m* for miles, the fare equation is written as follows:

$$f = \$1.25m + \$1.25$$

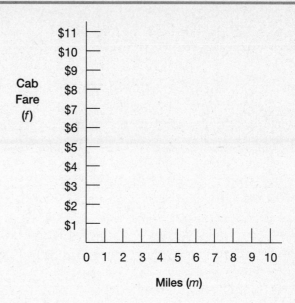

a) Using the fare equation, determine the fare of a cab trip of 8 miles.

b) Fill in the Table of Values for the *m* values listed, then graph the fare equation.

Table of Values

m	f
1	
3	
5	

c) Using the graph, estimate the distance a fare of $10.00 will take you.

4. In the United States, we use both the **Fahrenheit**-scale thermometer and the **Celsius**-scale (metric) thermometer. The following equation gives the Fahrenheit temperature (°F) for a given Celsius temperature (°C).

$$°F = \tfrac{9}{5}°C + 32°$$

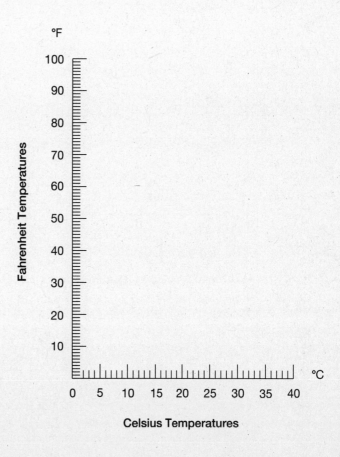

a) Water boils at 100°C. Using the temperature equation, determine the temperature in °F of the boiling point of water.

b) Fill in the Table of Values for the °C values listed; then graph the temperature equation.

Table of Values

°C	°F
15	
25	
35	

c) Water freezes at 32°F. From the graph, estimate the equivalent Celsius temperature.

Finding the Slope

You are already familiar with the concept of **slope.** When you walk uphill, you walk up a slope. When you walk downhill, you walk down a slope. A graphed line, like a hill, has a slope.

The slope of a line is given as a number. When you move between two points on the line, the slope is found by dividing the change in y value by the change in x value.

$$\text{Slope of a line} = \frac{\text{Change in } y \text{ value}}{\text{Change in } x \text{ value}}$$

- A line that goes up from left to right has a *positive slope.*
- A line that goes down from left to right has a *negative slope.*

On Graph A, the slope of line A is 2.

On Graph B, the slope of line B is −1.

Graph A

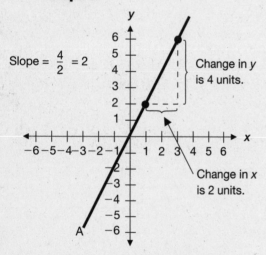

Slope $= \dfrac{4}{2} = 2$

Change in y is 4 units.

Change in x is 2 units.

Line A has *positive slope* because it goes *up* from left to right.

Graph B

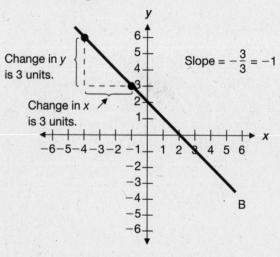

Change in y is 3 units.

Change in x is 3 units.

Slope $= -\dfrac{3}{3} = -1$

Line B has *negative slope* because it goes *down* from left to right.

0 Slope and Undefined Slope

Two types of lines have neither positive nor negative slope.
- A horizontal line has *zero slope.* The x axis is a line with 0 slope.
- A vertical line has an *undefined slope.* The concept of slope does not apply to a vertical line. The y axis is a line with undefined slope.

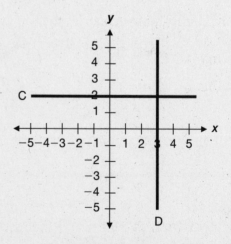

Line C has 0 slope.
Line D has undefined slope.

Practice

1. Name the slope of each line below as *positive, negative, zero,* or *undefined.*

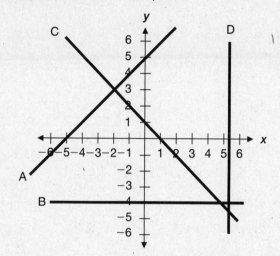

Line A: _____   Line C: _____

Line B: _____   Line D: _____

2. Find the numerical value of the slope of Line E graphed below.

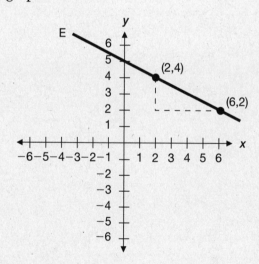

Slope of Line E is _____ .

Problem 3 refers to the following graph.

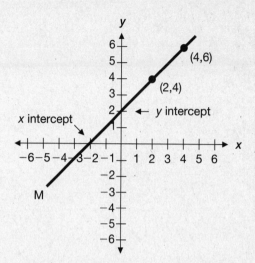

3. a) Find the numerical value of the slope of line M.

   b) What are the coordinates of the *x intercept,* the point where line M crosses the *x* axis?

   c) What are the coordinates of the *y intercept,* the point where line M crosses the *y* axis?

4. By subtracting coordinates, determine the slope of the line that passes through each of the following pair of points. Part a is completed as an example.

   a) (2,4) and (5,7)

   $$\text{slope} = \frac{\text{change in } y}{\text{change in } x} = \frac{7-4}{5-2} = \frac{3}{3} = 1$$

   b) (0,0) and (3,4)

   c) (2,5) and (6,8)

Test Readiness Checkup

Circle each correct answer. Check your answers on page 250; then correct any errors.

1. What rule do you follow to solve the following equation for x?

   $4x = 36$

   (1) Subtract 4 from each side.
   (2) Multiply each side by 4.
   (3) Subtract 4 from each side.
   (4) Divide each side by 4.
   (5) Divide each side by 36.

2. During a sale, Jan paid 75% of the original price of a sweater. If Jan paid $27 for the sweater, which equation below can be solved to find the original price (p)?

   (1) $0.75 = \$27p$
   (2) $0.75p = \$27$
   (3) $0.25p = \$27$
   (4) $0.25 = p - \$27$
   (5) $0.75 = p - \$27$

3. What is the solution of the following equation?

   $3x + 9 = 27$

   (1) $x = 4$ (4) $x = 7$
   (2) $x = 5$ (5) $x = 8$
   (3) $x = 6$

4. Which equation below represents the information given in the following statement?

   "Four times y minus 2 is equal to 2 times y plus 6."

   (1) $4y - 2 = 2y + 6$
   (2) $4y + 2 = 2y + 6$
   (3) $4y - 2 = 2y - 6$
   (4) $4y - 2 = 2y - 6$
   (5) $4y + 2y = 6 - 2$

5. If you subtract 8 from 7 times a certain number, you end up with 34. What is the number?

   (1) 4
   (2) 5
   (3) 6
   (4) 7
   (5) 8

6. Karen and Robin share monthly expenses. However, Robin pays $50 less each month than Karen. How much is Karen's share of January's $780 total?

   (1) $365
   (2) $415
   (3) $465
   (4) $545
   (5) $730

7. Frank, Albert, and Carmine worked together on a job. Frank earned $40 less than Albert. Carmine earned twice as much as Frank. If the three men together earned $380 for the job, how much did Frank earn?

   (1) $55   (4) $105
   (2) $70   (5) $125
   (3) $85

8. **Consecutive even numbers** are even whole numbers that follow one another: for example, 4, 6, and 8. If the sum of consecutive even numbers is 42, which equation can be used to find the smallest of the three numbers (x)?

   (1) $x + 2x + 4x = 42$
   (2) $x + 2x + 4x + 2 = 42$
   (3) $x + x + x = 42$
   (4) $x + x + 2 + x + 4 = 42$
   (5) $x + 2 + 4 = 42$

9. What is the missing term (value of y) in the following proportion?

$$\frac{12}{9} = \frac{48}{y}$$

(1) 36
(2) 42
(3) 57
(4) 60
(5) 64

10. Elvira walked 3 miles in 40 minutes. If she maintains this rate, about how many minutes will it take Elvira to walk 7 miles?

(1) 64
(2) 79
(3) 93
(4) 107
(5) 124

11. In a local poll of 98 shoppers, 5 out of every 7 said they prefer low-fat milk over whole milk. Which proportion can be used to find the number of shoppers who said they prefer low-fat?

(1) $\frac{7}{12} = \frac{x}{98}$

(2) $\frac{5}{7} = \frac{98}{x}$

(3) $\frac{5}{12} = \frac{x}{98}$

(4) $\frac{7}{12} = \frac{98}{x}$

(5) $\frac{5}{7} = \frac{x}{98}$

12. In the equation $y = 4x - 7$, what is the value of y when $x = 2$?

(1) 0
(2) 1
(3) 2
(4) 3
(5) 4

Problem 13 refers to the following diagram.

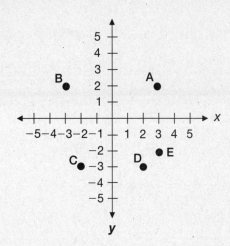

13. Which of the lettered points on the graph above has the coordinates (-3,2)?

(1) A   (4) D
(2) B   (5) E
(3) C

Problem 14 refers to the diagram below.

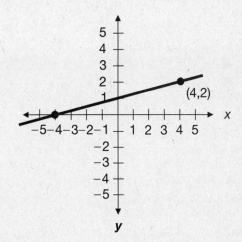

14. The line on the graph contains the point (4,2). Knowing this, determine which of the following is the equation of this line.

(1) $y = 2x + 1$

(2) $y = \frac{1}{3}x + 2$

(3) $y = \frac{1}{2}x - 2$

(4) $y = 3x - 1$

(5) $y = \frac{1}{4}x + 1$

UNIT 9 Geometry

Geometry includes the study of angles and triangles and the study of perimeter, area, and volume.

Introducing Angles

An **angle** is formed when two straight lines meet at a point.

- The lines that form the angle are called the **sides** of the angle.
- The point where the lines meet is called the **vertex** of the angle.

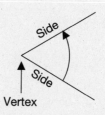

The symbol for angle is ∠.

An arc ⌒ is usually drawn to show which of two possible angles is meant.

For example, ∠ might represent either ∠ or ⌒.

Labeling Angles

An angle is labeled (named) in one of the three ways shown below.

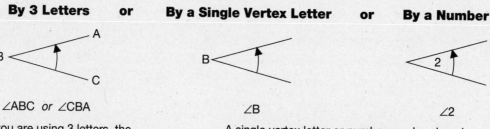

By 3 Letters or	**By a Single Vertex Letter** or	**By a Number**
∠ABC *or* ∠CBA	∠B	∠2
When you are using 3 letters, the vertex letter is always written as the second letter.	A single vertex letter or number can be placed either outside or inside the angle.	

Practice

Write a letter or number name for each angle pictured below.

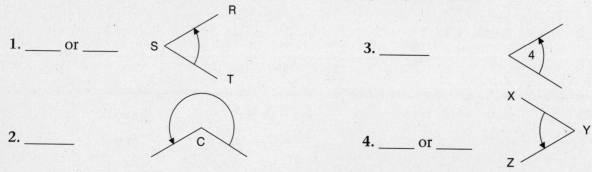

1. _____ or _____

2. _____

3. _____

4. _____ or _____

Measuring Angles

The size of an angle depends only on the opening between its sides. This opening is measured in units called **degrees**. The symbol for degrees is °. A 45-degree angle is written as 45°.

Degrees can be thought of as parts of a divided circle. A whole circle contains 360°.

180° is $\frac{1}{2}$ of a circle.

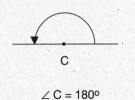

$\angle C = 180°$

90° is $\frac{1}{4}$ of a circle.

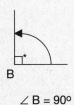

$\angle B = 90°$

*A small square □ is often used to indicate a 90° angle.

1° is $\frac{1}{360}$ of a circle.

$\angle A = 1°$

Example

A whole pie is cut into 8 equal pieces. What angle do the sides of each piece make?
A whole pie is a circle. To find the number of degrees in each piece, divide 360° by 8.
Each piece = $\frac{360°}{8} = 45°$.

Answer: Each piece is cut at a **45° angle.**

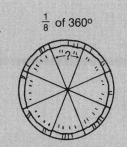

$\frac{1}{8}$ of 360°

A whole circle contains 360°.

Practice

■ Solve.

1. Suppose you want to construct a circle graph that contains 6 equal segments.

   a) At what angle should you draw the sides of each segment?

   b) Draw the angle identified in part a.

2. A spinner for a new board game is equally likely to stop in any one of 10 same-size sections.

   a) At what angle do the sides of each section meet?

   b) Draw the angle identified in part a.

■ Match each angle with its approximate size.

_____ 3.   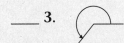   a) 135°

_____ 4.   b) 25°

_____ 5.   c) 230°

_____ 6.   d) 75°

_____ 7.   e) 180°

Angle Relationships

Types of Angles

An angle can also be classified by its size.

Acute Angle

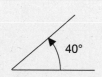

40°

More than 0° but less than 90°

Right Angle

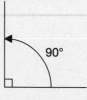

90°

Exactly 90°

Obtuse Angle

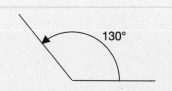

130°

More than 90° but less than 180°

Straight Angle

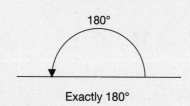

180°

Exactly 180°

Reflex Angle

200°

More than 180° but less than 360°

Pairs of Angles

Angles can be added to form a single larger angle.

Two angles that add to 90° are called **complementary angles.**

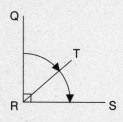

∠ QRT + ∠ TRS = 90°

∠ QRT and ∠ TRS are complementary
∠ QRT is the *complement* of ∠ TRS.

Two angles that add to 180° are called **supplementary angles.**

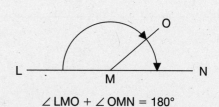

∠ LMO + ∠ OMN = 180°

∠ LMO and ∠ OMN are supplementary.
∠ LMO is the *supplement* of ∠ OMN.

Example

As shown at right, ∠ ABC is a right angle (90°).
What is ∠ ABD if ∠ DBC is 47°?
∠ ABD = 90° − 47° = 43°

∠ ABD and ∠ DBC are
complementary angles:
∠ ABD + ∠ DBC = 90°

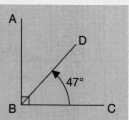

Answer: ∠ ABD = 43°

Practice

■ Name each angle below: *acute, right, obtuse, straight,* or *reflex.*

1. _____

29°

2. _____

300°

3. _____

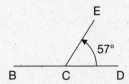
140°

4. _____

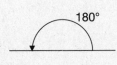
180°

5. _____

90°

6. _____

68°

■ Find the value of each angle indicated below.

7. ∠ DEG = _____

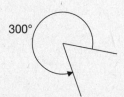
D
G
37°
E   F

8. ∠ BCE = _____

E
57°
B   C   D

9. ∠ PNO = _____

M
O   52°
P   N

■ Solve.

10. The diagonal brace in a fence gate makes a 38° angle with the bottom crosspiece. What angle does the brace make with the vertical side piece?

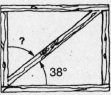
?
38°

12. A ladder is leaning against the side of a house. The ladder makes an acute angle of 64° with the ground. What is the value of the obtuse angle the ladder makes with the ground?

64° ?

11. The side of an A-frame house makes a 125° angle with the ground. How large is the acute angle that this side makes with the floor of the house?

125°
?

13. The two sides of a picture frame are cut so that they can be joined at the ends. At what angle must each side be cut so that the angle made by the joined sides is 90°?

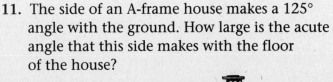
?
?

Angles Formed by Intersecting Lines

Many angles are formed by lines crossing one another. Here are two examples that are often used as the basis of test questions.

Vertical Angles

Vertical angles are the angles opposite each other when two straight lines cross.

Vertical angles are equal.

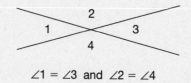

$\angle 1 = \angle 3$ and $\angle 2 = \angle 4$

When two lines meet (or cross) at a right angle, the lines are called **perpendicular lines**. Vertical angles formed by crossing perpendicular lines are all 90° angles. The symbol for perpendicular is ⊥. At right, line A is perpendicular to line B, or A ⊥ B.

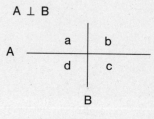

$\angle a = \angle b = \angle c = \angle d = 90°$

Parallel Lines Cut by a Transversal

Parallel lines are straight lines that run side by side and never cross. The symbol for parallel is ∥. At right, line C is parallel to line D, or C ∥ D.
A **transversal** is a third line that crosses two parallel lines.

When two parallel lines are cut by a transversal
• the four acute angles are equal
• the four obtuse angles are equal
• each acute angle is supplementary to each obtuse angle

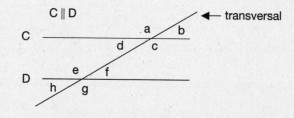

$\angle b = \angle d = \angle f = \angle h$
$\angle a = \angle c = \angle e = \angle g$

and

$\angle a + \angle b = 180°$
$\angle a + \angle d = 180°$
$\angle b + \angle e = 180°$
and so on.

Practice

■ **Find the value of each angle indicated below.**

1. ∠ C =

2. ∠ 2 =

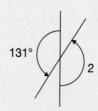

3. ∠ x =

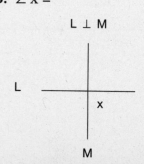

Solve the following problems.

4. $X \parallel Y$

$\angle b =$ _____

$\angle c =$ _____

$\angle d =$ _____

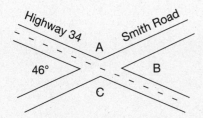

5. $A \parallel B$

$\angle 1 =$ _____

$\angle 2 =$ _____

$\angle 3 =$ _____

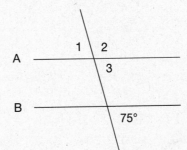

6. $S \parallel T$

$\angle x =$ _____

$\angle y =$ _____

$\angle z =$ _____

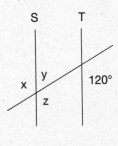

7. Smith Road intersects Highway 34 as shown below. What are the values of the three unmeasured angles?

$\angle A =$ _____

$\angle B =$ _____

$\angle C =$ _____

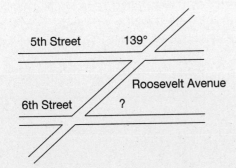

9. The capital letter E is formed by joining parallel lines with a transversal. Name four other capital letters that are formed with parallel lines and a transversal.

E

8. Roosevelt Avenue crosses 5th and 6th streets as shown below. If 5th Street is parallel to 6th Street, what is the value of the acute angle that Roosevelt makes with 6th Street?

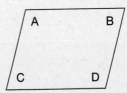

10. A **parallelogram** is a four-sided figure that has two pairs of parallel lines. Using your knowledge of angles, find the sum of the four angles of a parallelogram.
(Hint: Each obtuse angle is supplementary to each acute angle.)

$\angle A + \angle B + \angle C + \angle D =$ _____

Working with Triangles

A triangle is a closed plane (flat) figure that has 3 sides and 3 angles. Triangles are named with 3 letters, one placed at the vertex of each angle. The symbol for triangle is △.

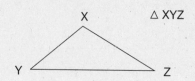

Types of Triangles

Here are the 4 most common types of triangles.

Equilateral Triangle

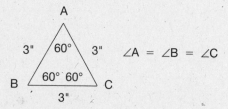

An **equilateral triangle** has three equal sides and three equal, 60° angles.

Isosceles Triangle

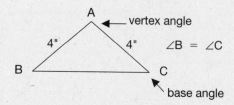

An **isosceles triangle** has 2 equal sides and 2 equal angles called **base angles**. The 3rd angle is called the **vertex angle**.

Scalene Triangle

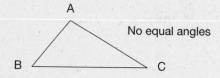

A **scalene triangle** has no equal sides and no equal angles.

Right Triangle

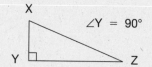

An isosceles or scalene triangle that has a 90° angle is also called a **right** triangle.

Angle Relationships in a Triangle

> The sum of the 3 angles in a triangle is equal to 180°.

What is the value of ∠B in ∠ABC?

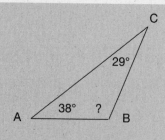

Step 1. Add ∠A and ∠C.
∠A + ∠C = 67°

Step 2. To find ∠B, subtract 67° from 180°.
∠B = 180° − 67° = 113°

Answer: ∠B = 113º

∠A + ∠B + ∠C = 180°

Practice

▨ **Name each triangle below: *equilateral*, *isosceles*, or *scalene*. Circle any *right triangle*.**

1. _____

43°
43°

2. _____

90° 35°

3. _____

2m   2m

2m

4. _____

3"
7"
$9\frac{1}{2}$"

▨ **Find the value of each angle indicated below.**

5. ∠ B = _____

B
?
A 39°   62° C

6. ∠ G = _____

F
21°
? 27° H
G

7. ∠ Y = _____

X ___ Y
40° ?
26°
Z

▨ **Solve.**

8. What is the roof angle (angle x) in the drawing below?

38° 38°

9. Referring to the drawing below, how large is the acute angle that the support makes with the shelf?

90°
47° **SUPPORT**

10. Look carefully at the drawing below and answer the following questions.

a) At what acute angle does the Amtrak line cross Maple Street?

b) How large is the acute angle that Smith's property line makes with the Amtrak line?

(Hint: As a first step, determine the value of ∠ a.)

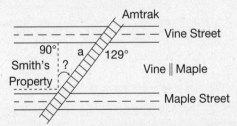

Amtrak
Vine Street
90° a 129°
Smith's ?
Property
Vine ∥ Maple
Maple Street

▨ **Use algebra to solve problems 11–12.**

11. ∠ A and ∠ B are supplementary angles. ∠ A is 4 times as large as ∠ B. Determine the size of each angle.

(Hint: Let ∠ A = 4*x* and ∠ B = *x*. Let the sum = 180°; then solve.)

12. In Δ CDE, ∠ C is the smallest angle. ∠ D is twice as large as ∠ C, and ∠ E is three times as large as ∠ C. Determine the size of ∠ C, ∠ D, and ∠ E.

(Hint: Let ∠ C = *x*, ∠ D = 2*x*, and ∠ E = 3*x*. Let the sum = 180°; then solve.)

Problem Solver: Solving Similar Triangles

Similar triangles are triangles where each of the angles in one triangle equals an angle in the other triangle. Similar triangles have the same shape and differ only in the lengths of their sides. The symbol ~ stands for *is similar to*.

   In similar triangles, the sides that are opposite the equal angles are called **corresponding sides**. The lengths of corresponding sides can be written as a proportion.

Example 1

△ ABC and △ DEF are similar triangles because they have the same three angles. We can write a proportion as follows:

Step 1. Identify the pairs of corresponding sides.
   a) AC and DF are corresponding sides.
      (Each is opposite an 80º angle.)
   b) BC and EF are corresponding sides.
      (Each is opposite a 40º angle.)
   c) AB and DE are corresponding sides.
      (Each is opposite a 60º angle.)

Step 2. Write a proportion using two of the pairs.

$$\frac{BC}{EF} = \frac{AC}{DF} \text{ or } \frac{10}{15} = \frac{16}{24}$$

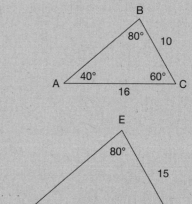

Note: The numerator of each fraction is from one triangle, while each denominator is from the other triangle.

Example 2 shows how to find an unknown length when working with similar triangles.

Example 2

△ XYZ and △ RST are similar triangles. The length of side YZ is unknown and is represented as *d*. Find this length.

Step 1. Write a proportion using the corresponding sides. To simplify your work, write the proportion so that *d* (the unknown length) is the numerator of the left-side fraction.

$$\frac{YZ}{ST} = \frac{XZ}{RT} \text{ or } \frac{d}{32} = \frac{30}{40}$$

Step 2. Solve the proportion for *d*.

$$d = 32(\frac{30}{40}) = \overset{8}{\cancel{32}}(\frac{3}{\cancel{4}_1}) = 24$$

Answer: The length of YZ is 24.

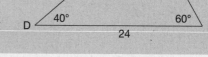

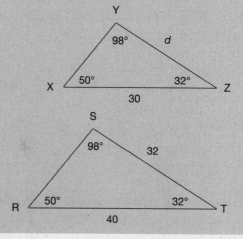

Practice

■ **Write and solve a proportion to find each unknown length.**

1. △ ABC ~ △ DEF. What is length *d*?

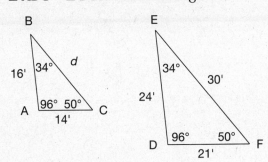

2. △ HIJ ~ △ RST. What is length *x*?

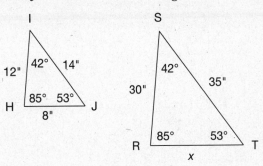

■ **Use a proportion to solve each problem below.**

3. Two trees stand side by side. The smaller tree is 12 feet high and casts a shadow of 20 feet. At the same time of day, the larger tree casts a shadow of 55 feet. How tall is the larger tree?

(Hint: Similar triangles are formed by the trees, the ground, and the dotted lines that represent the sun's rays.)

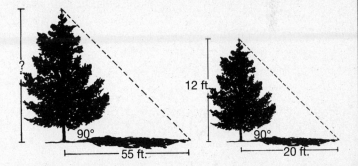

4. At the point of her campsite, Aida estimated the distance (*d*) across Odell River. To do this, she walked off the distances as shown in her drawing at right. Use her drawing to estimate the distance across the river.

(Hint: The two triangles are similar because all three angles are equal. Here's why:

a) Each has a right angle (90º). These angles are indicated by squares.

b) The vertical angles, labeled 1 and 2, are equal.

c) The third angles, labeled 3 and 4, are also equal. Reason: If two triangles have two equal angles, the third angles are equal because the sum of the angles in each triangle must be 180º.)

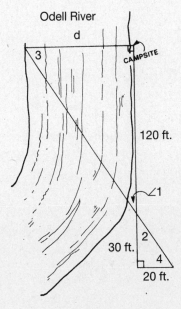

Problem Solver: Solving Similar Figures

Proportions can also be used to solve problems involving other types of similar figures. Any two figures are similar if they have equal angles and if their dimensions (lengths, widths, heights, etc.) are in proportion.

Enlarging or Reducing a Picture

You produce a similar figure when you enlarge or reduce a photograph or drawing.

Example 1

Marsha wants to enlarge a photo that measures 3 inches high by 5 inches wide so that the copy is 12 inches wide.
What will be the height of the copy?

Step 1. Letting h be the height of the copy, write a proportion as follows:

$$\frac{\text{height of copy}}{\text{height of original}} = \frac{\text{width of copy}}{\text{width of original}}$$

or $\frac{h}{3} = \frac{12}{5}$

Step 2. Solve for h.

$$h = \frac{3 \times 12}{5} = \frac{36}{5} = 7\frac{1}{5} \text{ inches}$$

Answer: $7\frac{1}{5}$ **inches**

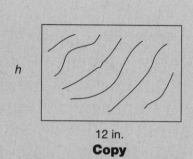

12 in.
Copy

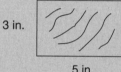

3 in.

5 in.
Original

Working with a Scale Drawing

Scale drawings, such as blueprints and maps, are drawings that are *similar* to the represented object. The **scale** tells the ratio of the corresponding dimensions. For example, if a scale is "1 inch = 24 inches," the ratio of the actual object to the drawing is 24 to 1.

Example 2

Hal is looking at his house blueprint. The scale reads "1 inch = 18 inches." What is the actual width of a hallway that measures $2\frac{1}{4}$ inches wide on the blueprint?

To find the width of the hallway, multiply the drawing width by 18.
Hallway width = $18 \times 2\frac{1}{4} = 40\frac{1}{2}$ inches

Answer: $40\frac{1}{2}$ **inches**

Example 3

A map of the United States has a scale that reads "1 inch = 250 miles." What is the distance by airplane of two cities whose map distance is $6\frac{1}{2}$ inches?

To find the distance by air, multiply the map distance by 250.
Air distance = $250 \times 6\frac{1}{2} = 1,625$ miles

Answer: 1,625 miles

Practice

■ **Solve each problem.**

1. As shown below, Connie is going to have a photograph enlarged. The larger copy is to be 18 inches wide. What will be the height of this copy?

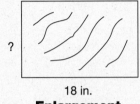

4 in.

6 in.
Original Photo

?

18 in.
Enlargement

4. David wants to reduce a photograph to wallet size. At most, the smaller copy can be 4 inches high. At this height, what will be the width of the wallet-size copy?

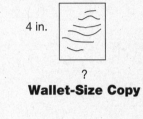

10 in.

8 in.
Original Photograph

4 in.

?
Wallet-Size Copy

2. On a building blueprint, a room measures $6\frac{1}{2}$ inches wide by 9 inches long. The blueprint scale reads "1 inch = 2 feet." What are the actual dimensions of this room?

width: _____ length: _____

5. On a United States map having a scale that reads "1 inch = 300 miles," Kelly measures the straight-line distance from Miami to Boston as $4\frac{1}{8}$ inches. What is the approximate air distance between these two cities?

3. As shown below, a photograph that measures 7 inches high by 10 inches wide is placed in a frame. Each side of the frame is 1 inch wide.

   a) What are the outer dimensions of the frame?

   height: _____ width: _____

   b) Is the rectangle formed by the outer edge of the frame similar to the rectangular photograph?

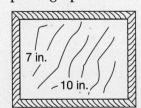

7 in.

10 in.

6. By air, New York and Chicago are about 660 miles apart. How many inches (to the nearest quarter inch) will these cities be apart on a map on which the scale reads "1 inch = 200 miles"?

(Hint: To find map distance, divide the actual distance by the map scale ratio.)

Understanding Squares and Square Roots

Squares

The **square** of a number is that number multiplied by itself. The square of 7 is $7 \times 7 = 49$. In symbols, we write the square of a number as a **base** and an **exponent**.

$$7 \times 7 \text{ is written } 7^2 \leftarrow \text{ exponent}$$
$$\uparrow \text{ base}$$

The exponent (2) tells how many times to write the base (7) as a **factor** (a number being multiplied). The exponent of a square is always 2. Read 7^2 as "7 squared" or "7 to the 2nd power." The *value* of 7^2 is 49. Here are three more examples:

Squared Number	As a base and an exponent	Read in words	Value
3×3	3^2	"three squared"	9
10×10	10^2	"ten squared"	100
$b \times b$	b^2	"b squared"	*

*A value for "variable squared" can be found only when you know the value of the variable. If $b = 4$, then $b^2 = 16$; if $b = 9$, then $b^2 = 81$.

Square Roots

To find the **square root** of a number, you ask, "What number times itself equals this number?"
   What number times itself equals 49? The answer is 7: $7 \times 7 = 49$.
   The symbol for square root is $\sqrt{}$. We write $7 = \sqrt{49}$.

The table below contains the squares of numbers from 1 to 15. These numbers 1, 4, 9, 16, and so on, are called **perfect squares** because their square roots are whole numbers.

Table of Perfect Squares

$1^2 = 1$	$6^2 = 36$	$11^2 = 121$
$2^2 = 4$	$7^2 = 49$	$12^2 = 144$
$3^2 = 9$	$8^2 = 64$	$13^2 = 169$
$4^2 = 16$	$9^2 = 81$	$14^2 = 196$
$5^2 = 25$	$10^2 = 100$	$15^2 = 225$

The values in this table are used to find the square roots of these perfect squares. For example, if $c^2 = 121$, what is c? To determine c, find 121 on the table.

   The table shows $11^2 = 121$. Thus, $11 = \sqrt{121}$, and $c = \textbf{11}$.

Practice

■ **Find the value of each squared number or letter below.**

1. $5^2 =$

2. $8^2 =$

3. $10^2 =$

4. $(\frac{1}{2})^2 =$

5. $(3.5)^2 =$

6. $(\frac{2}{3})^2 =$

7. If $a = 6$, $a^2 =$

8. If $x = 3.2$, $x^2 =$

9. If $y = \frac{7}{8}$, $y^2 =$

■ **Use the Table of Perfect Squares to find each square root below.**

10. $\sqrt{16} =$

11. $\sqrt{100} =$

12. $\sqrt{36} =$

13. $\sqrt{169} =$

14. $\sqrt{225} =$

15. $\sqrt{196} =$

16. If $a^2 = 25$, $a =$

17. If $b^2 = 49$, $b =$

18. If $x^2 = 121$, $x =$

■ **Find each value below.** *Find the value of each squared number before finding a sum or difference.*

19. If $a = 5$ and $b = 6$, $a^2 + b^2 =$

20. If $x = 7$ and $y = 4$, $x^2 - y^2 =$

21. If $a = 3$ and $b = 4$, $\sqrt{a^2 + b^2} =$

22. If $m = 10$ and $n = 8$, $\sqrt{m^2 - n^2} =$

■ **Estimate the value of each square root below. Two are done as examples. Then use a calculator to find a more exact value.***

23. $\sqrt{50} \approx 7$

Since $50 \approx 49$. Calculator answer: ≈ 7.07

24. $\sqrt{35} \approx$

25. $\sqrt{90} \approx 9.5$

Since 90 is about halfway between 81 and 100. Calculator answer: ≈ 9.49

26. $\sqrt{12} \approx$

27. $\sqrt{98} \approx$

28. $\sqrt{155} \approx$

*To find a square root with a calculator, enter the number and then press the $\boxed{\sqrt{}}$ (or $\boxed{\sqrt{x}}$) key.

Learning the Pythagorean Theorem

In a *right triangle*, the side opposite the right angle (90°) is called the **hypotenuse.** The hypotenuse is the longest side.

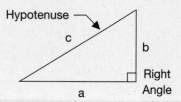

The Greek mathematician Pythagoras discovered an important relationship between the hypotenuse and the two shorter sides. This relationship is called the ***Pythagorean theorem.***

In words: In a right triangle, the square of the hypotenuse is equal to the sum of the squares of the other two sides.

In symbols: $c^2 = a^2 + b^2$ (using the labels on the triangle above)

If you know the lengths of the two shorter sides, you can use the Pythagorean theorem to find the length of the hypotenuse.

What is the length of the hypotenuse of $\triangle$ LMN?

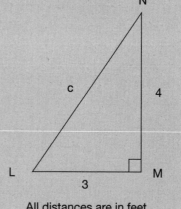

Step 1. Find the square of each shorter side.
a) Substitute 3 for a and find a^2.
$$a^2 = 3^2 = 9$$
b) Substitute 4 for b and find b^2.
$$b^2 = 4^2 = 16$$

Step 2. Add the squares found in Step 1.
$$c^2 = a^2 + b^2$$
$$= 9 + 16$$
$$= 25$$

Step 3. To find c, take the square root of 25.
$$c = \sqrt{25} = 5$$

All distances are in feet.

Answer: The length of the hypotenuse is 5 feet.

You can also find the length of a side of a right triangle if you know the lengths of the hypotenuse and the other side. To do this, rewrite the Pythagorean theorem as follows:

In symbols: $a^2 = c^2 - b^2$

In words: *Side a squared* equals *hypotenuse squared* minus *side b squared*.

In the example above, if you know that the hypotenuse is 5 and that one side is 4, you can find the other side by subtracting squares:
$$a^2 = 5^2 - 4^2$$
$$= 25 - 16$$
$$= 9$$

or, $a = \sqrt{9} = 3$ (as we already know)

Practice

Using the Pythagorean theorem, find the length of each unmeasured side below. (For problems 1–5, refer to the Table of Perfect Squares on page 212.)

1. $c =$

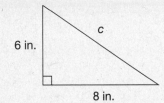

6 in.

c

8 in.

2. $x =$

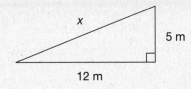

x

5 m

12 m

3. $d =$

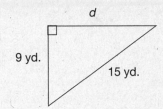

d

9 yd.

15 yd.

Solve each problem below.

4. A **rectangle** is a four-sided figure with four right angles. A **diagonal** divides a rectangle into two right triangles. What is the length of the diagonal in the rectangle below?

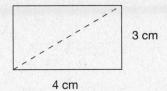

3 cm

4 cm

5. Jenny hiked 12 miles north and several miles east during the first day of her camping trip. If she is now 15 miles from her car, how many miles east has she hiked?

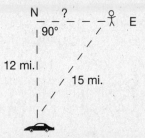

N ? E
90°
12 mi.
15 mi.

For problems 6–7, refer to the Partial Table of Perfect Squares shown below.

Partial Table of Perfect Squares			
$15^2 = 225$	$25^2 = 625$	$31^2 = 961$	$39^2 = 1,521$
$16^2 = 256$	$26^2 = 676$	$32^2 = 1,024$	$40^2 = 1,600$
$17^2 = 289$	$27^2 = 729$	$33^2 = 1,089$	$41^2 = 1,681$

6. A ladder rests against the side of Kate's house. The bottom of the ladder is 8 feet from the house, and the top just reaches a window that's 15 feet above ground. How long is the ladder?

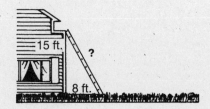

15 ft.
?
8 ft.

7. The Coast Guard determines that a disabled ship is 41 miles from the station. The ship is directly east of Summit Peak, 26 miles directly north. To the nearest mile, how far is the ship from shore?

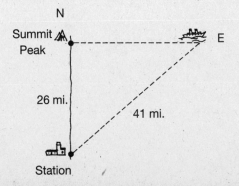

N
Summit Peak
E
26 mi.
41 mi.
Station

Number Highlight: More About Exponents

Powers: Positive and Negative Exponents

As you've seen, you can use the exponent 2 to write the square of a number.

You can also use larger **positive exponents** as a shorthand way to indicate more **factors** (numbers being multiplied). When reading this shorthand, you refer to the exponent as the **power**. *The power tells how many times to write the base as a factor when you multiply.*

Examples	As a base and an exponent	Read in words	Value
$5 \times 5 \times 5$	5^3	"5 to the 3rd power"*	125
$6 \times 6 \times 6 \times 6$	6^4	"6 to the 4th power"	1,296
$10 \times 10 \times 10 \times 10 \times 10$	10^5	"10 to the 5th power"	100,000

*5^3 is also called "5 cubed."

A **negative exponent** represents the number found by *inverting* the number represented by a positive exponent. Compare the examples below with the examples above.

Examples	As a base and an exponent	Read in words	Value
$\dfrac{1}{5 \times 5 \times 5}$	5^{-3}	"5 to the minus 3rd power"	$\dfrac{1}{125}$
$\dfrac{1}{6 \times 6 \times 6 \times 6}$	6^{-4}	"6 to the minus 4th power"	$\dfrac{1}{1,296}$
$\dfrac{1}{10 \times 10 \times 10 \times 10 \times 10}$	10^{-5}	"10 to the minus 5th power"	$\dfrac{1}{100,000}$ *or* 0.00001

Practice

▮ Complete the following table.

	As a base and an exponent	Read in words	Value
1. $2 \times 2 \times 2$	_____	_____	____
2. $10 \times 10 \times 10 \times 10$	_____	_____	____
3. $\dfrac{1}{3 \times 3 \times 3}$	_____	_____	____
4. $\dfrac{1}{10 \times 10 \times 10 \times 10}$	_____	_____	____

▮ Find each value.

5. $7^3 =$ 　　　6. $2^{-5} =$ 　　　7. $10^6 =$ 　　　8. $10^{-3} =$

Scientific Notation

Scientific notation is a way to write numbers that contain many zeros. In scientific notation, a number is written as the product of two factors: (1) a number between 1 and 10 *and* (2) a power of 10.

The power tells how many places you move the decimal point to write the number using digits.
- A positive exponent tells you to move the decimal point to the right.
- A negative exponent tells you to move the decimal point to the left.

Positive Exponents		Negative Exponents	
Example: $8 \times 10^4 = 80,000$		Example: $3 \times 10^{-4} = 0.0003$	
4 decimal places right		4 decimal places left	
Number	In Scientific Notation	Number	In Scientific Notation
90,000,000	9×10^7	0.00006	6×10^{-5}
570,000	5.7×10^5	0.00085	8.5×10^{-4}
753,000,000	7.53×10^8	0.000096	9.6×10^{-5}

Practice

Using a positive exponent, write each number in scientific notation.

9. $4,000 =$

10. $8,000,000 =$

11. $9,000,000,000 =$

12. $25,000 =$

13. $75,000,000 =$

14. $1,380,000 =$

Using a negative exponent, write each number in scientific notation.

15. $0.006 =$

16. $0.0002 =$

17. $0.000008 =$

18. $0.00075 =$

19. $0.0000925 =$

20. $0.00000318 =$

Solve.

21. The distance from the earth to the moon is approximately 240,000 miles. Write this distance in scientific notation.

22. The distance from the earth to the sun is about 9.3×10^7 miles. Write this distance as a whole number.

23. Using a microscope, a biologist measured the diameter of a pollen grain to be 0.038 cm. Write this diameter in scientific notation.

24. The diameter of a hydrogen atom is about 2×10^{-8} cm. Write this diameter as a decimal.

Finding the Perimeter

The distance around a plane (flat) object is known as its **perimeter.** You find the perimeter of a many-sided object by adding the lengths of its sides. The symbol for perimeter is P. For squares, rectangles, and triangles, you often see the **perimeter formulas** written below.

Perimeters of Squares, Rectangles, and Triangles

Shape	Main Features	Formula for Perimeter
Square 6 ft.	Four right angles Four equal sides	$P = 4s$ where s = side

$P = 4 \times 6$ (equal to P = 6 + 6 + 6 + 6)
    = 24 feet

Shape	Main Features	Formula for Perimeter
Rectangle 4 m 7 m	Four right angles Two pairs of equal sides	$P = 2l + 2w$ where l = length w = width

$P = 2 \times 7 + 2 \times 4$ (equal to P = 7 + 7 + 4 + 4)
    = 14 + 8
    = 22 meters

Shape	Main Features	Formula for Perimeter

Triangle 3 yd. 2.5 yd. 6 yd.	Three angles
	Three sides
		$P = s_1 + s_2 + s_3$ where s_1 = side 1
		                  s_2 = side 2
		                  s_3 = side 3

$P = 6 + 3 + 2.5$
    = 11.5 yards

Practice

Find the perimeter of each figure below.

1. P = _____

2. P = _____

3. P = _____

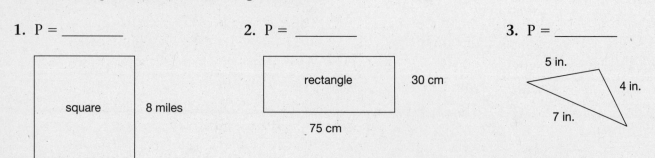

square   8 miles

rectangle   30 cm   75 cm

5 in.   4 in.   7 in.

Finding the Circumference of a Circle

A **circle** is a figure in which all the points are at an equal distance from the center. This distance is called the **radius.** The symbol for radius is *r*.

The distance across a circle is called the **diameter.** As you may know, the diameter passes through the center of the circle and is equal to twice the radius. The symbol for diameter is *d*.

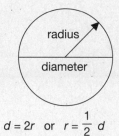

$$d = 2r \quad \text{or} \quad r = \frac{1}{2}d$$

The distance around a circle is called the **circumference.** The symbol for circumference is C.

The circumference is given by the formula $C = \pi d$, where π has the approximate value 3.14 or $\frac{22}{7}$ (π is known as pi and is pronounced "pie").

> To compute the circumference of a circle, multiply π times the diameter.

Example 1	**Example 2**
What is the circumference of a circle that has a diameter of 7 feet?	Find the circumference of a circle that has a diameter of 2.5 yards.
To find the circumference, multiply $\frac{22}{7}$ by 7.	Circumference = 3.14×2.5
$C = \frac{22}{7} \times 7 = \frac{22}{7} \times \frac{7}{1} = 22$	$C = 3.14 \times 2.5 = 7.85$
Answer: 22 feet	**Answer: 7.85 yards**
Note: Use $\pi \approx \frac{22}{7}$ when the radius or diameter is a multiple of 7: 7, 14, 21, 28, and so on.	Note: Use $\pi \approx 3.14$ when the radius or diameter is given as a decimal.

Practice

Using the formula $C = \pi d$, solve each problem below.

1. The Community Wading Pool is in the shape of a circle. If the radius of the pool is 28 feet, what is the distance around the pool?

2. The tire on Frank's bicycle has a radius of $12\frac{1}{4}$ inches.

   a) What is the diameter of this tire?

   b) To the nearest inch, what is the tire's circumference?

3. The diameter of a circular garden is 6.5 meters. To the nearest tenth meter, what is the circumference of this garden?

4. A sandbox in the shape of a circle has a radius of 3.3 yards.

   a) What is the diameter of the sandbox?

   b) To the nearest yard, what is the circumference of this sandbox?

Finding the Area of Squares and Rectangles

Area is a measure of surface. For example, to measure the size of a floor, you determine its area. The symbol for area is **A**.

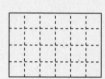

In the English system, the most common area units are the **square inch** (sq. in.), **square foot** (sq. ft.), and **square yard** (sq. yd.). In the metric system, the most common area units are the **square centimeter** (cm^2) and the **square meter** (m^2).

1 foot

1 foot

The area of a square is given by the formula $A = s^2$ ($s \times s$), where *s* stands for *side*.

> To compute the area of a square, multiply the side by itself.

A = 9 area units

The area of a rectangle is given by the formula $A = lw$ ($l \times w$) where *l* stands for *length* and *w* stands for *width*.

> To compute the area of a rectangle, multiply length by width.

A = 24 area units

Example 1
What is the area of a square that measures 3 feet on each side?

3 ft.

3 ft.

To compute this area, multiply 3×3.

$A = 3 \times 3 = 9$

Answer: 9 square feet

Example 2
What is the area of a rectangle that is 11 ft. long and 4 ft. wide?

4 ft.

11 ft.

To find this area, multiply 11×4.

Area $= 11 \times 4 = 44$

Answer: 44 square feet

Practice

▌ Find the area of each figure.

1. A = _____

9 m

9 m

2. A = _____

10 in.

7 in.

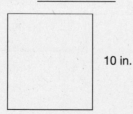

3. A = _____

8 cm

14 cm

Solve each problem below. You may find it useful to draw a picture.

4. Linda is buying a piece of glass to fit inside a square picture frame. How many square inches of glass does she need if the space in which the glass fits measures 7.5 inches by 7.5 inches?

5. The area of the top of a square table is 16 square feet. How long is each side of the table?
   (Hint: For a square, $s = \sqrt{A}$)

6. Using square tiles that measure 9 inches on each side, Ben is tiling his shop floor. The square floor is 12 feet long and 12 feet wide.

   a) What is the area in square feet of each tile? (Hint: 9 in. = $\frac{3}{4}$ ft.)

   b) What is the area of the shop floor?

   c) How many tiles will Ben need to cover this floor?

7. Brent's bedroom floor measures $4\frac{1}{2}$ yards long and $3\frac{3}{4}$ yards wide. How many square yards of carpet will he need to buy to cover this floor?

8. The label on a gallon of interior wall paint reads: "Coverage 400 square feet." What length of 8-foot-high wall can be painted with one gallon of this paint?
   (Hint: For a rectangle $l = A \div w$.)

9. Using square bricks that measure 8 inches on each side, Ellen is building a retaining wall that measures 24 feet long and 4 feet high.

   a) What is the area in square feet of each brick? (Hint: 8 in. = $\frac{2}{3}$ ft.)

   b) What is the area of the wall?

   c) How many bricks will Ellen need to build this wall?

Two-Step Area Problems

To solve a two-step area problem, divide a complex figure into shapes you're familiar with; then solve for the area of each shape separately.

Find the area of each room pictured below.

10. A = _____

    (Hint: Think of the room as being made up of two squares.)

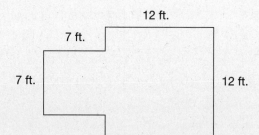

11. A = _____

    (Hint: Think of the room as being made up of a square and a rectangle.)

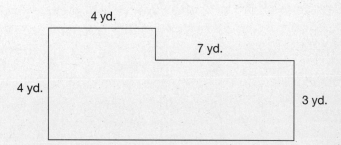

Finding the Area of a Triangle

The area of a *triangle* is given by the formula $A = \frac{1}{2}bh$, where *b* stands for *base* and *h* stands for *height*.

The **base** is one side of the triangle. The **height** is the distance from the base to the vertex of the opposite angle. Only in a right triangle is the height one of the sides of the triangle.

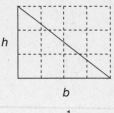

$$A = \frac{1}{2}bh$$

> To compute the area of a triangle, multiply $\frac{1}{2}$ by the base by the height.

Example 1	Example 2	Example 3
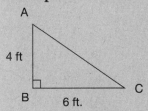	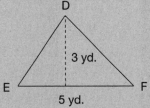	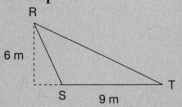
The height (4) is a side of right triangle ABC.	The height (3) is drawn as a dotted line within $\triangle$ DEF.	The height (6) is drawn as a dotted line outside of $\triangle$ RST.
$A = \frac{1}{2}bh$ $= \frac{1}{2} \times \overset{3}{\cancel{6}} \times 4$ $= 12$ square feet	$A = \frac{1}{2}bh$ $= \frac{1}{2} \times 5 \times 3$ $= 7\frac{1}{2}$ square yards	$A = \frac{1}{2}bh$ $= \frac{1}{2} \times 9 \times \overset{3}{\cancel{6}}$ $= 27$ square meters

Practice

▮ **Find the area of each triangle drawn below.**

1. A = _____

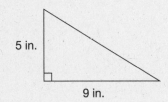

2. A = _____

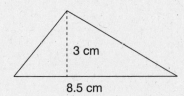

3. A = _____

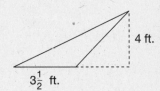

▮ **Solve each problem below.**

4. Katie's garden is in the shape of a triangle. What is the area of this garden if the base measures 18 yards and the height measures 10 yards?

5. Cal is being charged $1.50 per square foot of plywood. If he buys a triangular piece that measures 8 feet long (the base) by 3 feet high (the height), how much will he be charged?

Finding the Area of a Circle

The area of a circle is given by the formula $A = \pi r^2$ ($\pi \times r \times r$) where $\pi \approx 3.14$ or $\pi \approx \frac{22}{7}$, and r = radius.

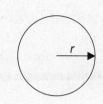

To find the area of a circle, multiply π by the radius by the radius.

Example

A circular wading pool has a radius of 14 feet. Find the area of this pool.

Write $\frac{22}{7}$ for π and 14 for r in the area formula. Use cancellation if possible.

$$A = \pi \times r \times r = \frac{22}{7} \times \overset{2}{\cancel{14}} \times 14 = 616$$

Answer: 616 square feet

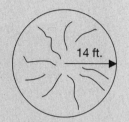

Wading Pool

Practice

▪ **Find the approximate area of each circle below.**

1. A = _____

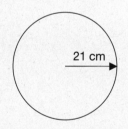

21 cm

2. A = _____

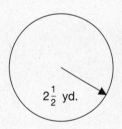

$2\frac{1}{2}$ yd.

3. A = _____

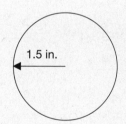

1.5 in.

▪ **Solve each problem below.**

4. Radio station WPSB broadcasts its signal to all points within a 50-mile radius of the station. What is the approximate area served by this station?

5. A sprinkler waters in a circular pattern, spraying water out 9.5 feet in all directions. To the nearest square foot, what area of lawn does this sprinkler water?

6. A custom-made circular mirror is costing Lou 10¢ per square inch. How much will Lou be charged for a mirror that has a radius of 16 inches?

7. The ratio of the radiuses of two circles is 3 to 1. What is the ratio of their areas?

(Hint: Represent the radius of the larger circle as 3 and the smaller as 1. The number π cancels out when you write the ratio of the two areas.)

Finding the Volume of a Rectangular Solid

Volume is a measure of the space taken up by a solid object such as a rock or is a measure of the space enclosed by a box or other surface. The symbol for volume is **V**.

In the English system, the most common **volume units** are the **cubic inch** (cu. in.), **cubic foot** (cu. ft.), and **cubic yard** (cu. yd.). In the metric system, the most common volume units are the **cubic centimeter** (cm^3) and the **cubic meter** (m^3).

The most common shape you'll ever find the volume of is a **rectangular solid**—familiar shapes such as boxes, suitcases, freezers, and rooms.

Sample Volume Unit

1 cubic yard

Rectangular Solid

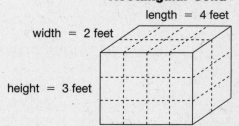

> To find the volume of a rectangular solid, multiply the length by the width by the height. In symbols: **V = lwh** ($l \times w \times h$)

Volume = $l \times w \times h$
= $4 \times 2 \times 3$
= 24 cubic feet

Practice

■ **Find the volume of each rectangular solid pictured below.**

1. V = _____

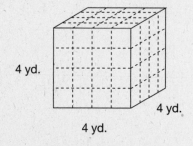

4 yd.

4 yd.

4 yd.

A cube is a rectangular solid in which $l = w = h$.

2. V = _____

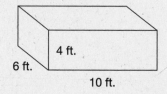

4 ft.

6 ft.

10 ft.

3. V = _____

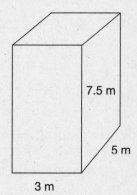

7.5 m

5 m

3 m

■ **Solve each problem.**

4. A water bed measures 6 feet long, 5 feet wide, and $\frac{1}{2}$ foot high.
  a) How many cubic feet of water does this water bed hold?
  b) If water weighs 62 pounds per cubic foot, what is the weight of water in this bed when full?

5. Billy has two play blocks, each in the shape of a cube. The larger block measures 6 inches on each side. The smaller block measures 2 inches on each side. What is the ratio of the volume of the larger block to the volume of the smaller block?

Finding the Volume of a Cylinder

The volume of a cylinder is given by the formula

$$V = \pi r^2 h \ (\pi \times r \times r \times h)$$

where $\pi \approx 3.14$ or $\pi \approx \frac{22}{7}$, r = radius, and h = height.

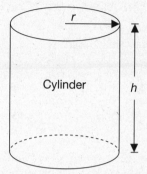

Cylinder

> To find the volume of a cylinder, multiply the area of its bottom (the circle with area πr^2) by its height h.

The top and bottom of a cylinder are equal size circles.

Example

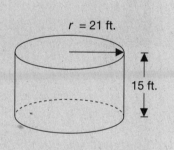

A cylindrical water tower has a radius of 21 feet and a height of 15 feet. How many cubic feet of water does this tower hold?

Write $\frac{22}{7}$ for π, 21 for r, and 15 for h.
$$V = \pi \times r \times r \times h = \frac{22}{7} \times 21 \times 21 \times 15 = 20,790$$

Answer: 20,790 cubic feet

r = 21 ft.

15 ft.

Practice

▌ **Find the *approximate* volume of each cylinder pictured below.**

1. V = _____

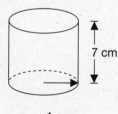

7 cm

r = 1 cm

2. V = _____

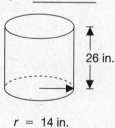

26 in.

r = 14 in.

3. V = _____

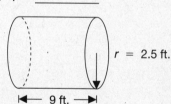

r = 2.5 ft.

9 ft.

▌ **Solve each problem.**

4. The cylindrical tank of a gasoline truck is 25 feet long and has a radius of 3.5 feet.
   a) To the nearest cubic foot, what is the volume of this tank?
   b) If 1 cubic foot ≈ 7.5 gallons, about how many gallons of gas can the truck carry each load?

5. Two cylinders are the same height. However, the ratio of the radiuses of the cylinders is 3 to 1. Find the ratio of their volumes.
   (Hint: Represent the radius of the larger cylinder as 3 and the smaller as 1. The numbers π and h cancel out when you write the ratio of the volumes.)

Test Readiness Checkup

Circle each correct answer. Check your answers on page 252; then correct any errors.

1. Francine cut a large circular cake into 12 equal slices. What acute angle is formed by the sides of each slice?

   (1) 25°
   (2) 30°
   (3) 45°
   (4) 90°
   (5) 120°

2. What name is given to an angle that measures 124°?

   (1) acute   **(4)** straight
   (2) right   **(5)** reflex
   (3) obtuse

3. ∠ A is *supplementary* to ∠ B. What is the sum of ∠ A plus ∠ B?

   (1) 45°   **(4)** 180°
   (2) 60°   **(5)** 360°
   (3) 90°

Problem 4 refers to the drawing below.

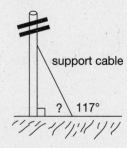

4. What is the value of the acute angle that the telephone pole's support cable makes with the ground?

   (1) 37°   **(4)** 92°
   (2) 63°   **(5)** 114°
   (3) 79°

Problems 5–6 refer to the drawing below.

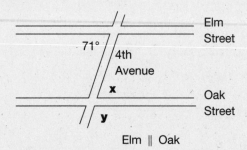

Elm ‖ Oak

5. What is the value of the acute angle labeled ∠ x above?

   (1) 19°   **(4)** 64°
   (2) 26°   **(5)** 71°
   (3) 53°

6. What is the value of the obtuse angle (∠ y) that Oak Street makes with 4th Avenue?

   (1) 81°   **(4)** 161°
   (2) 109°   **(5)** 171°
   (3) 149°

Problem 7 refers to the triangle below.

vertex angle

50°

3 ft.   3 ft.

7. The isosceles triangle above has a vertex angle of 50°. How large is each of the two equal base angles?

   (1) 40°   **(4)** 95°
   (2) 55°   **(5)** 130°
   (3) 65°

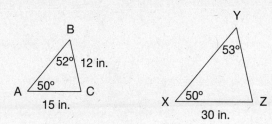

8. △ ABC is similar to △ XYZ. What is the length of side YZ?

(1) 24 in.   (4) 32 in.
(2) 26 in.   (5) 35 in.
(3) 30 in.

9. On a map that has a scale that reads "1 inch = 250 miles," Jamie measures the distance between Newport and Edison as $2\frac{1}{4}$ inches. What is the approximate distance between these cities?

(1) 400 miles
(2) 450 miles
(3) 500 miles
(4) 550 miles
(5) 600 miles

10. What is the best estimate of the square root of 72?

(1) 6.5
(2) 7
(3) 7.5
(4) 8
(5) 8.5

11. The two shorter sides of a right triangle measure 6 feet and 8 feet long. What is the length of the hypotenuse?

(1) 2 feet   (4) 28 feet
(2) 10 feet   (5) 48 feet
(3) 14 feet

12. The rose garden at Central Park is in the shape of a circle. Passing through the center, the distance across the garden is 42 feet. Using $\pi = \frac{22}{7}$, determine the distance around the garden.

(1) 20 feet   (4) 121 feet
(2) 67 feet   (5) 132 feet
(3) 94 feet

13. Matt bought a circular rug to place on the floor of his den. The floor of the den is 18 feet long and 15 feet wide. If the radius of the rug is 7 feet, about what percent of the den floor will the rug cover?

(1) 24%   (4) 68%
(2) 41%   (5) 79%
(3) 57%

14. How many cubic feet of storage are in a freezer that has inside dimensions of 4 feet by 2.25 feet by 2.5 feet?

(1) 14   (4) 20.25
(2) 16.5   (5) 22.5
(3) 18.75

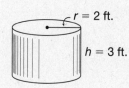

15. Using the volume formula $V = \pi r^2 h$, determine the volume of the cylindrical drum to the nearest cubic foot.

(1) 38 cu. ft.   (4) 50 cu. ft.
(2) 43 cu. ft.   (5) 56 cu. ft.
(3) 46 cu. ft.

Post-Test

Solve the following problems.

1. For her salon, Gina bought 11 bottles of shampoo on sale for $2.79 per bottle. Which of the following is the *best estimate* of the change Gina will receive if she pays with two $20 bills?

   (1) $5
   (2) $10
   (3) $15
   (4) $20
   (5) $25

Problem 2 refers to the following list.

Family Burgers Customer Count	
Day	Customers
Monday	689
Tuesday	714
Wednesday	733
Thursday	673

2. *Estimate* the total number of customers who visited Family Burgers during the four days shown above.

   (1) 2,000
   (2) 2,400
   (3) 2,800
   (4) 3,200
   (5) 3,600

3. Which of the following inequalities is *not* true?

   (1) $-5 > -3$
   (2) $0 > -2$
   (3) $-4 < -1$
   (4) $3 > -2$
   (5) $7 > 0$

Problem 4 refers to the following number line.

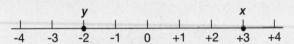

4. Using values represented on the number line, what is the difference $x - y$?

   (1) -5
   (2) -1
   (3) 0
   (4) 1
   (5) 5

5. Playing on the game show "You Keep the Money," Todd Jackson had a balance of $-$350 in winnings after the 2nd round. In the 3rd round, though, Todd won $750. What was Todd's balance after round 3?

   (1) $400
   (2) $600
   (3) $800
   (4) $1,000
   (5) $1,100

6. Bruce purchased a case of car oil that was marked down $2.00 below its normal price of $12.40. Bruce was also given a rebate coupon for $1.50. Which equation represents the cost (c) to Bruce for the case of oil after he received the rebate?

   (1) $c = \$12.40 - \$2.00 + \$1.50$
   (2) $c = \$12.40 - \$2.00 - \$1.50$
   (3) $c = \$12.40 + \$2.00 - \$1.50$
   (4) $c = \$12.40 + \$2.00 + \$1.50$
   (5) Not enough information is given.

7. The average price of a new home in Portland is $119,231. What is this amount rounded to the nearest $1,000?

   (1) $100,000
   (2) $110,000
   (3) $119,000
   (4) $120,000
   (5) $125,000

8. Two hundred twenty yards is what fraction of a mile? (1 mile = 5,280 feet)

(1) $\frac{1}{24}$

(2) $\frac{1}{16}$

(3) $\frac{1}{10}$

(4) $\frac{1}{8}$

(5) $\frac{1}{4}$

9. A cup (8 fluid ounces) is filled with 6 fluid ounces of milk. Which of the following does *not* represent the part of the cup that is filled?

(1) $\frac{6}{8}$

(2) 75%

(3) $\frac{3}{4}$

(4) 0.75

(5) 68%

Problem 10 refers to the following list.

200-Meter Race Times
Carlo: 24.07 sec.
Young-Soo: 23.8 sec.
Kenny: 24.1 sec.
Pat: 23.795 sec.

10. Listing the winner first, in what order did the four runners listed above finish?

(1) Young-Soo, Pat, Kenny, Carlo
(2) Pat, Young-Soo, Carlo, Kenny
(3) Carlo, Kenny, Young-Soo, Pat
(4) Kenny, Carlo, Young-Soo, Pat
(5) Young-Soo, Kenny, Carlo, Pat

11. Andre's boss asked him to arrange the following bolts in a storage drawer according to length, placing the shortest bolt near the front.

$1\frac{3}{4}$ in. $1\frac{5}{8}$ in. $1\frac{11}{16}$ in.

Bolt A   Bolt B   Bolt C

Listing the shortest bolt first, what is the correct order of bolts?

(1) A, B, C
(2) A, C, B
(3) B, A, C
(4) B, C, A
(5) C, B, A

Problem 12 refers to the drawing below.

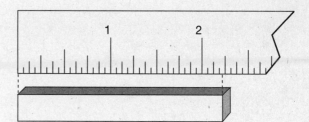

12. To the nearest $\frac{1}{8}$ inch, what is the width of the board shown above?

(1) 2 in.

(2) $2\frac{1}{8}$ in.

(3) $2\frac{1}{4}$ in.

(4) $2\frac{3}{8}$ in.

(5) $2\frac{1}{2}$ in.

13. During winter-league play, Larry had scores of 150 or more in 73% of the games he bowled. In about what fraction of his games did Larry score 150 or more?

(1) $\frac{1}{2}$

(2) $\frac{5}{8}$

(3) $\frac{2}{3}$

(4) $\frac{3}{4}$

(5) $\frac{7}{8}$

Anita places the following ingredients in an 8-cup mixing bowl: $\frac{7}{8}$ cup of white flour, $1\frac{3}{4}$ cups of whole-wheat flour, $\frac{1}{2}$ cup of wheat bran, and $\frac{7}{8}$ cup of dark rye flour.

14. At this point, what fraction of the mixing bowl has Anita filled?

    (1) $\frac{1}{8}$

    (2) $\frac{1}{4}$

    (3) $\frac{3}{8}$

    (4) $\frac{1}{2}$

    (5) $\frac{5}{8}$

15. In Anita's mixture, what is the ratio of the amount of whole-wheat flour to the amount of white flour?

    (1) $\frac{1}{2}$

    (2) $\frac{4}{7}$

    (3) $\frac{1}{1}$

    (4) $\frac{3}{2}$

    (5) $\frac{2}{1}$

16. Working 5 days a week, 8 hours a day, Ravi earns $288 a week. Next week, though, Ravi plans to work only 24 hours. For this work, he'll receive this same hourly rate of pay. Which expression below tells how much Ravi will earn next week?

    (1) $\frac{3}{5}$ ($288)

    (2) $\frac{2}{5}$ ($288)

    (3) $\frac{2}{3}$ ($288)

    (4) $288 - \frac{2}{3}$ ($288)

    (5) $288 - \frac{3}{5}$ ($288)

17. Natalie's bathtub holds 80 gallons of water. To check the drainage rate, she measured the time it takes for the tub to empty. If the full tub empties in 5 minutes and 20 seconds, at what rate in gallons per minute does Natalie's tub drain?

    (1) $12\frac{1}{2}$ gal. per min.

    (2) 15 gal. per min.

    (3) $18\frac{1}{3}$ gal. per min.

    (4) 21 gal. per min.

    (5) $24\frac{2}{3}$ gal. per min.

NMP Average Monthly Stock Value		
March	April	May
$38\frac{1}{4}$	$37\frac{3}{8}$	$36\frac{1}{2}$

18. The average monthly value of NMP Company stock for a 3-month period is shown above. If this trend continues, what will be NMP's average monthly stock value for June?

    (1) $35\frac{3}{4}$

    (2) $35\frac{5}{8}$

    (3) $35\frac{1}{2}$

    (4) $35\frac{3}{8}$

    (5) $35\frac{1}{4}$

19. During the week, Lanna drove her car a total of 268 miles on 13.8 gallons of gas. To the nearest tenth of a mile per gallon, how many miles per gallon is Lanna's car averaging?

    (1) 19.4
    (2) 19.9
    (3) 20.3
    (4) 20.8
    (5) 21.2

20. The Changs' heating bill for January is $329. During January, they used 4,000 kilowatt-hours of electric power. To the nearest *tenth of a cent*, what are the Changs being charged per kilowatt-hour?

(1) $0.80
(2) $0.08
(3) $0.082
(4) $0.083
(5) $8.3

21. Aurora receives a 6% real estate commission for each house she sells. What will be Aurora's commission on a house that sells for $84,000?

(1) $504
(2) $1,540
(3) $4,840
(4) $5,040
(5) $50,400

22. A washing machine that sold for $319.99 last year is priced at $399.89 this year. Which expression below gives the best *estimate* of the *percent of price increase* of this washing machine during the past year?

(1) $\frac{\$320}{\$400} \times 100\%$
(2) $\frac{\$400}{\$320} \times 100\%$
(3) $\frac{\$400-\$320}{\$320} \times 100\%$
(4) $\frac{\$400-\$320}{\$400} \times 100\%$
(5) $\frac{\$320+\$400}{\$320} \times 100\%$

23. To purchase a new pickup truck, Chen must make a 20% down payment. If he is required to put down $3,250, what is the price of the truck?

(1) $12,920
(2) $13,540
(3) $14,760
(4) $15,840
(5) $16,250

24. In a public opinion poll, 119 of the 548 people surveyed said they *do not* plan to buy an American car as their next car. Which expression gives the percent of those surveyed who *do* plan to buy an American car as their next car?

(1) $\frac{119}{548} \times 100\%$
(2) $\frac{548}{119} \times 100\%$
(3) $\frac{548-119}{119} \times 100\%$
(4) $\frac{548-119}{548} \times 100\%$
(5) $\frac{119}{548-119} \times 100\%$

Problems 25–26 refer to the following label.

GOODGROW PLANT FOOD	
Net Wt. 5 lb. (2.27 kg)	
Ingredients:	
Phosphorus	12%
Potassium	8%
Nitrogen	4%
Calcium	4%
Magnesium	2%

25. What total percent of ingredients in a 5-lb. box of Goodgrow Plant Food is not written on the label shown above?

(1) 30%
(2) 40%
(3) 50%
(4) 60%
(5) 70%

26. In a 5-lb. box of Goodgrow Plant Food, *about* what total weight is made up of phosphorus and calcium? (1 pound = 16 ounces)

(1) 9 ounces
(2) 13 ounces
(3) 1 pound 4 ounces
(4) 1 pound 11 ounces
(5) 2 pounds 7 ounces

27. Suppose you deposit $750 in a savings account that pays 4.9% simple interest. Which expression below gives the best *estimate* of the total amount that will be in your account at the end of 2 years and 1 month?

(1) $750 + ($750 × 0.05 × 2)
(2) $750 − ($750 × 0.5 × 2)
(3) $750 + ($750 × 0.5 × 2)
(4) $750 − ($750 × 0.05 × 2)
(5) $750 × 0.05 × 2

■ **Problems 28–29 refer to the graph below.**

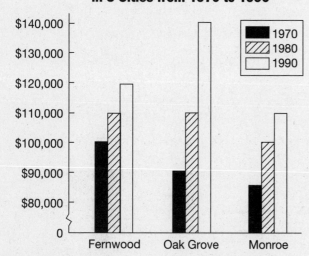

Average Price of New Houses in 3 Cities from 1970 to 1990

28. In which town and decade did average new-house prices increase the most?

(1) in Fernwood between 1970 and 1980
(2) in Oak Grove between 1970 and 1980
(3) in Monroe between 1970 and 1980
(4) in Fernwood between 1980 and 1990
(5) in Oak Grove between 1980 and 1990

29. By *about* what percent did average new-house prices increase in Monroe between 1980 and 1990?

(1) 10%
(2) 20%
(3) 30%
(4) 40%
(5) 50%

30. The 3 numbers of Ruth's combination lock are 8, 19, and 31, though she can't remember which order they're in! What's the probability that the correct combination is 19, 31, 8?

(1) $\frac{1}{9}$
(2) $\frac{1}{6}$
(3) $\frac{1}{5}$
(4) $\frac{1}{3}$
(5) $\frac{1}{2}$

■ **Problems 31–32 are based on the circle graph.**

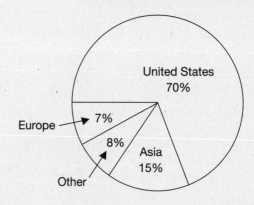

Lewis College Place of Origin of Students

31. Of the next 80 students who register at Lewis College, how many most likely will be from Asia?

(1) 12
(2) 15
(3) 18
(4) 21
(5) 24

32. *Estimate* the probability that the next two students who register at Lewis College will both be from the United States.

(1) 30%
(2) 40%
(3) 50%
(4) 60%
(5) 70%

33. What computation do you do to solve the following equation?

$$\frac{4}{5}y = 12$$

(1) multiply each side by $\frac{4}{5}$

(2) multiply each side by $\frac{5}{4}$

(3) subtract $\frac{4}{5}$ from each side

(4) divide each side by $\frac{5}{4}$

(5) add $\frac{5}{4}$ to each side

34. At a Memorial Day sale, Lita bought a sweater marked "20% off." If Lita paid $24 for the sweater, which equation below can be solved to find the original price (p)?

(1) $p = \frac{1}{5}(\$24)$

(2) $p = p + \frac{1}{5}(\$24)$

(3) $p - \frac{1}{5}p + \$24 = 0$

(4) $p - \frac{1}{5}p = \$24$

(5) $p + \frac{1}{5}p = \$24$

35. Which one or more of the three values of x listed below is a solution for the equation $x^2 - 5x + 6 = 0$?

A. $x = 1$ B. $x = 2$ C. $x = 3$

(1) A only

(2) B only

(3) C only

(4) A and B only

(5) B and C only

36. Alex and Joel have a yard-care service. Because Alex owns the truck and lawn-care equipment, he earns $25 more each day than Joel does. On Saturday, the two men earned a total of $165. Which equation below can be solved to determine Joel's share (x) of Saturday's earnings?

(1) $x - x + \$25 = \$165$

(2) $x + x + \$25 = \$165$

(3) $x + x = \$165 + \$25$

(4) $x - \$25 = \$165$

(5) $x + \$25 = \$165$

37. Seven out of the first 230 VCRs tested were found to have a tracking problem. Seven hundred seventy more VCRs are going to be tested. Which proportion can be used to estimate the number (n) of these remaining VCRs that will likely have a similar problem?

(1) $\frac{7}{230} = \frac{1000}{n}$

(2) $\frac{7}{230} = \frac{n}{1000}$

(3) $\frac{7}{230} = \frac{770}{n}$

(4) $\frac{7}{230} = \frac{n}{770}$

(5) $\frac{230}{770} = \frac{n}{1000}$

38. In the equation $y = 3x - 2$, what is the value of y when $x = 4$?

(1) 2

(2) 6

(3) 10

(4) 12

(5) 14

39. Emerald Car Rental charges $19 a day plus $0.15 per mile for rental of its compact cars. Letting t stand for total cost and n for number of days of rental, which equation below represents the total cost of renting a compact car from Emerald?

(1) $t = \$19 + \$0.15n$

(2) $t = \$19 - \$0.15n$

(3) $t = \$19n + \$0.15$

(4) $t = \$19n - \$0.15$

(5) Not enough information is given.

Problem 40 refers to the following graph.

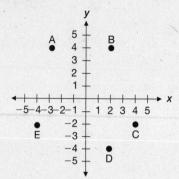

40. Which of the lettered points on the graph has the coordinates (4,−2)?

(1) A
(2) B
(3) C
(4) D
(5) E

41. When graphed, which of the following equations passes through the point (3,4)?

(1) $y = x - 3$
(2) $y = x + 1$
(3) $y = \frac{2}{3}x + 3$
(4) $y = \frac{1}{3}x + 1$
(5) $y = 2x - 1$

Problems 42–43 refer to the following graph.

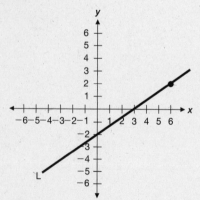

42. What is the slope of line L, graphed above?

(1) −1
(2) 0
(3) 1
(4) $\frac{2}{3}$
(5) $\frac{3}{2}$

43. What is the *y* intercept of line L?

(1) (0,−2)
(2) (−2,3)
(3) (−2,0)
(4) (3,−2)
(5) (3,0)

Problem 44 refers to the following drawing.

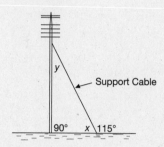

Support Cable

44. A support cable is used to help hold a radio broadcasting antenna in a vertical position. What is the value of the acute angle *y* that the cable makes with the antenna?

(1) 25º
(2) 65º
(3) 75º
(4) 80º
(5) 90º

Problem 45 refers to the following diagram.

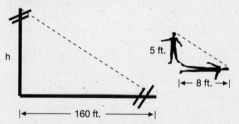

45. Gerald is standing near a telephone pole. Both he and the pole cast shadows on the level ground. Which proportion is the correct one to use to determine the height (*h*) of the pole?

(1) $\frac{3}{5} = \frac{160}{h}$
(2) $\frac{h}{5} = \frac{160}{h}$
(3) $\frac{3}{5} = \frac{h}{160}$
(4) $\frac{h}{8} = \frac{160}{5}$
(5) Not enough information is given.

Formulas for problems 46–48:
  Pythagorean theorum $c^2 = a^2 + b^2$
  Area of a triangle: $\frac{1}{2} bh$
  Area of a rectangle: lw
  Volume of a cylinder: $\pi r^2 h$

46. The two shorter sides of a right triangle measure 5 feet and 6 feet. What is the best *estimate* of the length of the hypotenuse?

(1) 4 ft.
(2) 5 ft.
(3) 6 ft.
(4) 7 ft.
(5) 8 ft.

■ **Problem 47 refers to the drawing below.**

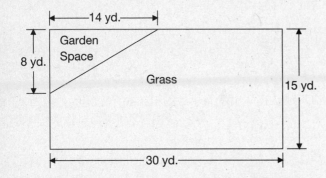

47. As shown above, the garden space in Shanda's backyard is in the shape of a triangle. The rest of the yard is planted in grass. How many square yards of grass are in this yard?

(1) 169
(2) 338
(3) 394
(4) 428
(5) 450

48. A cylindrical tank has a *diameter* of 3 feet and a height of 4 feet. Which expression below can be used to find the volume (V) of this tank in cubic feet?

(1) V = 3.14(4)(4)(3)
(2) V = 3.14(2)(2)(3)
(3) V = 3.14(3)(3)(4)
(4) V = 3.14(1.5)(1.5)(4)
(5) Not enough information is given.

49. A U.S. wall map has a scale of 1 inch = 125 miles. To the nearest 10 miles, what is the actual distance between two towns that are $8\frac{1}{4}$ inches apart on this map?

(1) 760 miles
(2) 980 miles
(3) 1,030 miles
(4) 1,190 miles
(5) 1,420 miles

50. Kari is planning to place a brick border around her backyard patio. The patio measures 27 feet long by 18 feet wide. If the bricks are 9 inches long, what is the fewest number of bricks that Kari needs for this border?

(1) 75
(2) 90
(3) 105
(4) 120
(5) 135

51. Which of the following expresses 1,875,400 in scientific notation?

(1) 18.754×10^5
(2) 1.8754×10^6
(3) 18.754×10^{-5}
(4) 1.8754×10^{-6}
(5) Not enough information is given.

52. Which equation does *not* have the same solution as the others?

(1) $2x + 2 = x + 4$
(2) $x + 8 = 12$
(3) $x \div 2 = 2$
(4) $2x = 14 - 6$
(5) $2x - 3 = x + 1$

53. As part of a fund-raising effort, Kim's class sold 225 raffle tickets for $4.00 each. The winner takes home a gas barbecue. Kim's parents bought 3 tickets. What is the probability that Kim's family will win the barbecue?

(1) $\frac{1}{225}$

(2) $\frac{1}{75}$

(3) $\frac{2}{75}$

(4) $\frac{4}{225}$

(5) $\frac{12}{225}$

54. Jan's car used 9.5 gallons of gas to drive between Rockford and Glendale. What more do you need to know to be able to determine what mileage Jan's car got on this trip?

(1) the price Jan paid per gallon of gas
(2) the highway mileage rating of Jan's car when new
(3) the length of time it took Jan to make the trip
(4) the distance between Rockford and Glendale
(5) the city mileage rating of Jan's car when new

■ **Problem 55 refers to the following diagram.**

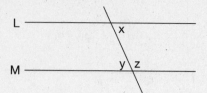

55. Which of the following do you need to know to find the value of $\angle x$?

A. the sum of $\angle y$ and $\angle z$

B. that line L is parallel to line M

C. the value of $\angle z$

(1) A only
(2) B only
(3) C only
(4) A and B
(5) B and C

56. Starting at the library on Jenson Avenue, Na walked 5 blocks east before turning south on 21st Street. She then walked 8 blocks, turned west onto Hilyard Avenue, and walked 12 more blocks to 33rd Street. How many blocks is Na from the library now if she can walk north on 33rd Street to Jenson and then turn east to go to the library?

(1) 12
(2) 15
(3) 18
(4) 25
(5) 31

Post-Test Answer Key

1. (2) $10
   $40 − ($3 × 10)
2. (3) 2,800
   about 700 × 4
3. (1) −5 > −3
4. (5) 5
   5 spaces separate −2 and 3
5. (1) $400
6. (2) $c = \$12.40 − \$2.00 − \$1.50$
7. (3) $119,000
8. (4) $\frac{1}{8}$
   220 yd. = 660 ft.; $\frac{660}{5,280} = \frac{1}{8}$
9. (5) 68%
10. (2) Pat, Young-Soo, Carlo, Kenny
11. (4) B, C, A
    $1\frac{5}{8} = 1\frac{10}{16}$, $1\frac{11}{16}$, $1\frac{3}{4} = 1\frac{12}{16}$
12. (3) $2\frac{1}{4}$ in.
    $2\frac{2}{8} = 2\frac{1}{4}$
13. (4) $\frac{3}{4}$
    $73\% \approx 75\% = \frac{3}{4}$
14. (4) $\frac{1}{2}$
    $\frac{7}{8} + 1\frac{3}{4} + \frac{1}{2} + \frac{7}{8} = 4$
15. (5) $\frac{2}{1}$
    $1\frac{3}{4} \div \frac{7}{8} = 2$
16. (1) $\frac{3}{5}(\$288)$
    $\frac{24}{40} = \frac{3}{5}$
17. (2) 15 gal. per. min.
    $80 \div 5\frac{1}{3} = 15$
18. (2) $35\frac{5}{8}$
    $36\frac{1}{2} − \frac{7}{8}$
19. (1) 19.4
20. (3) $0.082
    The same as 8.2¢, the 2 being in the tenths-of-a-cent-place
21. (4) $5,040
22. (3) $\frac{\$400−\$320}{\$320} \times 100\%$
23. (5) $16,250
24. (4) $\frac{548−119}{548} \times 100\%$
    548 − 119 people, out of 548, *do plan* to buy an American car.
25. (5) 70%
26. (2) 13 ounces
    Calcium and phosphorus make up 16% of the total; 16% of 5 lb. = 0.8 lb. ≈ 13 oz.
27. (1) $750 + ($750 × 0.05 × 2)
28. (5) in Oak Grove between 1980 and 1990
29. (1) 10%
30. (2) $\frac{1}{6}$
31. (1) 12
    15% of 80

32. (3) 50%
    0.7 × 0.7 = 0.49 ≈ 50%
33. (2) multiply each side by $\frac{5}{4}$
34. (4) $p − \frac{1}{5}p = \$24$
    $20\% = \frac{1}{5}$
35. (5) B and C only
36. (2) $x + x + \$25 = \$165$
    x = Joel's share, $x + \$25$ = Alex's share; the sum = $165
37. (4) $\frac{7}{230} = \frac{n}{770}$
38. (3) 10
    $y = 3(4) − 2$
39. (5) Not enough information is given.
    To find the total cost t, you must know the number of miles driven.
40. (3) C
41. (2) $y = x + 1$
    $4 = 3 + 1$
42. (4) $\frac{2}{3}$
    slope = $\frac{\text{change in } y}{\text{change in } x} = \frac{2−0}{6−3} = \frac{2}{3}$
43. (1) $(0, −2)$
    The line crosses the y axis at $(0, −2)$
44. (1) 25°
    First, notice that $\angle x = 65°$
    $(180° − 115°)$. Now, write
    $\angle y = 180° − 90° − 65° = 25°$.
45. (2) $\frac{h}{5} = \frac{160}{h}$
46. (5) 8 ft.
    hypotenuse = $\sqrt{25 + 36} = \sqrt{61} \approx 8$
47. (3) 394
    $(30 \times 15) − \frac{1}{2}(14 \times 8)$
48. (4) $V = 3.14 (1.5)(1.5)(4)$; $r = 1.5$
49. (3) 1,030 miles
50. (4) 120
    The patio has a perimeter of 90 feet, which is 1,080 inches. 1,080 ÷ 9 = 120.
51. (2) 1.8754×10^6
52. (1) $2x + 2 = x + 4$
    Each equation has the solution $x = 4$ except (1) where $x = 2$.
53. (2) $\frac{1}{75}$
    $\frac{3}{225} = \frac{1}{75}$
54. (4) the distance between Rockford and Glendale
55. (5) B and C
    If line L is parallel to line M, you know that the sum $\angle x + \angle z = 180°$, and you can determine $\angle x$ if you also know $\angle z$: $\angle x = 180° − \angle z$.
56. (2) 15
    Na must walk north for 8 blocks and then walk east for 7 blocks: 8 + 7 = 15.

Answer Key

Unit 1: Number Sense

Page 2
1. forty-six
2. three hundred eighty-one
3. one thousand, five hundred eight
4. seven thousand, sixty-two

Page 3
1. e
2. c
3. b
4. a
5. f
6. d
7. 61,872
8. 462,800
9. 5,835,000
10. seven thousand, nine hundred twenty-six
11. two hundred thirty-seven thousand, three hundred
12. ninety-two million, nine hundred thousand

Page 4
Part A.
1. <, =, >
2. <, >, <
3. =, <, >
4. <, >, <
Part B.
1st: $6,780
2nd: $6,779
3rd: $5,932
4th: $5,589
5th: $4,867
6th: $4,814

Page 5
1. T, T, F
2. F, F, F
3. T, F, F
4. T, F, T
5. 7, 6, any number other than 5
6. any number smaller than 5, any number larger than 20, any number larger than 32
7. any number smaller than 6, any number smaller than 12, any number smaller than 5
8. 6, any number smaller than 5, 8

Page 6
1. 40; 200; 1,000
2. 70; 800; 5,000

Page 6 (continued)
3. 40; 70; 20; 80; 100
4. 100; 400; 300; 400; 800
5. 3,000; 2,000; 7,000; 5,000; 6,000

Page 7
1. $2; $4; $12; $13
2. $20; $30; $110; $90
3. $500
4. $1,700
5. $4,000
6. $14,000

Page 8
In 1–3, any reasonable estimate is acceptable. Sample estimates are given here.

1. 90; 120; 500; 700
2. 1,900; 3,800; 5,000; 8,000
3. $13; $18; $190; $190
4. (2) 791
5. (3) $44.82
6. (1) 3,750

Page 9
In 1–3, any reasonable estimate is acceptable. Sample estimates are given here.

1. 20; 70; 40; 300
2. 500; 600; 2,000; 2,000
3. $5; $8; $40; $40
4. (2) 388
5. Estimating does not help us determine the correct answer.
   (1) $28.66
6. (3) 1,984

Page 10
In 1–3, any reasonable estimate is acceptable. Sample estimates are given here.

1. 320; 720; 1,200; 1,200
2. 2,000; 15,000; 30,000; 180,000
3. $64; $54; $600; $1,800
4. (2) 80 × 60
5. (3) 110 × 100
6. (1) $140 × 50

Page 11
1. a) 3,000; b) 12,000
2. a) $2,000; b) $8,000
3. a) $500; b) $2,500
4. 3,500 hamburgers
5. 8,000 miles

Pages 12–13

In 1–6, any reasonable estimate is acceptable. Sample estimates are given here.

1. 7; 6; 3; $4
2. 70; $30; 300; $80
3. $2,000; 300; $700; 900
4. 4; $3; 2; $4
5. 3; 10; $40; 30
6. 50; $200; 100; $200

Page 15

1. 470
2. 12
3. (a) cheeseburger
   (b) hamburger
   (c) cheeseburger
   (d) fish sandwich
4. (3) calories
5. (1) fat
6. $3,214
7. $3,154
8. $20,850 and $20,900
9. $20,900 and $20,950

Page 17

1. <, <, <, >
2. >, >, <, <
3. 2
4. 4
5. 6
6. −1
7. 3°F
8. 1,000 feet

Page 19

Answers may vary. The most common answers are given here.

1. a) 8
   b) 99,999,999
2. clear key C
3. "0." appears.
4. a) 0.58
   b) 8.45
   c) 239.
   d) 1956.

Pages 20–21

1. (5) two thousand, forty-three
2. (4) no ten thousands
3. (2) 3,500,460
4. (3) $6.45 < $7.12
5. (5) 7
6. (4) 3,300
7. (3) $370
8. (3) $2,000
9. (3) $6.00
          2.00
        +1.00
        ─────
10. (3) 6,030
11. (1) $526
12. (4) 2,058
13. (5) 213

14. (5) 110°F
15. (3) $1.15
16. (5) D, B, E, A, C

Unit 2: Whole Numbers

Pages 22–23

1. 57; 95; 67; 153; 134
2. 439; 509; 754; 919; 805
3. 3,388; 2,677; 5,198; 8,569; 9,250
4. $9.94; $7.65; $14.39; $14.09
5. 961; 1,641; 1,813; $2,612; 4,221
6. 304; $512; 1,341; 4,720; $8,498
7. 246; $652; 2,657
8. 2,183; $6,211; 21,336
9. 293 miles
10. about $380

Pages 24–25

1. 33; 31; 10; 111; 211
2. 182; 290; 285; 92; 480
3. $4.09; $4.62; $3.17; $2.28
4. $1.93; $3.52; $2.80; $6.60; $8.82
5. 376; 199; 58; 358; 359
6. 125; 172; 259; 1,055; 1,176
7. $2.25; $2.15; $4.46; $7.61; $23.52
8. 174; $3.46; 360
9. $89
10. 137 pounds
11. Estimate: $30; Exact Change: $30.21

Page 27

1. $25.00 − $17.98
2. 2,305 + 1,987
3. 62 + 97
4. $5.00−$1.98
5. $4.88 + $2.19
6. 1,345 − 277
7. (2) 53 − 47
8. (3) $105 + $86 + $75
9. (1) $a = $2,375 + $1,850$
10. (3) $t = 50¢ + 30¢ + 20¢$
11. (1) $200 − $60
12. (3) $m = 400 + 300$

Page 29

1. 115
2. 33 miles
3. 220 miles
4. (4) 79
5. $2.87
6. (3) $2.00 + $5.00
7. $0.84
8. (5) $3.95

Pages 30–31

1. 69; 280; 266; 357; 666
2. $34.50; $58.00; $62.50; $231.75
3. 850; 1,472; 2,491; 532; 5,135
4. $319.44; $194.81; 33,814; 45,750; 29,175
5. $128.60; $342.80; 44,340; 26,740
6. 522,720; 342,225; 121,176; 285,324
7. 4,819; 3,655; $141.75; $731.00
8. $24.13
9. 648 miles
10. Estimate: $300; Exact Price: $290.92

Pages 32–35

1. 9 r3; 4 r1; 7 r5; 6 r1; 7 r4
2. 54; 84; 59; 223; 242 r2
3. $2.75; $3.74; $1.85; $4.25
4. 70; 50; $70; 700; $7,000
5. $2.02; 302; $3.04; 302; 201
6. 204 r2; 109 r4; $1.04; $206; 302 r1
7. 3; 3; 4; 5; 7
8. 19; 22; 24; 46; 32
9. 9; 6; 8; 7 r6; 9 r5
10. 34 r1; 19; 39 r21; 21 r4; 15 r39
11. 110; 179; 237; 437 r17; 275 r 25
12. 74; 121; 117 r5; 183 r7
13. 13; 18 r20; 219 r6; 208 r18
14. 13
15. 68
16. 7
17. 9

Page 37

1. 123×17
2. $236 \div 9$
3. 748×2
4. $134.88 \div 6$
5. $3,540 \div 35$
6. 3
7. 72
8. 294
9. $10.06
10. 1,243
11. $7.15
12. 1,992
13. 484
14. 27

Page 39

1. $18.99 + ($5.50 \times 7)$
2. $\frac{(345 + 125)}{3}$
3. $45 - ($60 \div 2)$
4. $4(37 + 29 + 14)$
5. $100 \div (2 + 4)$
6. 32; 96; 30
7. 4; 30; 16
8. (3) $20 - ($10.98 + $4.89 + $0.88)$
9. (4) $\frac{$6.57 + $4.50 + $5.25}{3}$
10. (3) $64 \times 5 + 18×2,
    and (5) $2($64 + $18) + 3($64)$

Page 41

1. Mean: $21; Median: $20; Mode: $26
2. Mean: 1,640; Median: 1,550; Mode: 1,400
3. a) 26; b) 32; c) 27

4. a) A: $20; B: $16; C: $14; D: $15
   b) $16.25
   c) $15.50
5. (3) $h = (9 + 8 + 12 + 10) \div 4$

Pages 42–43

1. (3) $374
2. (4) 372
3. (4) 4,112
4. (1) $236.65
5. (3) 486
6. (2) 10
7. (5) $140
8. (5) $0.79 \times 12 - 1.50
9. (2) $\frac{(230 - 148)}{30}$
10. (3) 67
11. (1) $1,400 \times 6$
12. (3) $83,080
13. (5) $p = ($82,500 + $87,000) \div 2$

Unit 3: Numbers Smaller than 1

Pages 44–45

1. a) one-half
   b) one-third
   c) one-fourth
   d) one-fifth
   e) one-sixth
   f) one-seventh
   g) one-eighth
   h) one-ninth
   i) one-tenth
   j) one-hundredth
   k) one-thousandth

2. a) an hour half
   b) $14,500 ninety percent
   c) a pound three-fourths
   d) a mile four-tenths
   e) regular price thirty-five percent
   f) an inch eight-hundredths

3. a) d
   b) f
   c) p
   d) p
   e) d
   f) f

Pages 46–47

Notice that a single hyphen is used in writing decimal fractions: if the number part contains a hyphen, you do not place a hyphen to separate the number from the place value.

1. three-tenths
2. five-tenths
3. nine-hundredths
4. twenty-hundredths
5. seven-thousandths
6. eighty-thousandths
7. seventy-five thousandths

8. four hundred twenty-five thousandths
9. two dollars and five cents
10. twelve dollars and seven cents
11. three dollars and eighty-eight cents
12. forty dollars and fifty cents
13. A: 0.056; B: 0.3; C: 0.024; D: 0.209; E: 0.120; F: 0.07; G: 0.8
14. Lopez: 0.078; Jenkins: 0.2; Morris: 0.55; Myers: 0.112; Calder: 0.6; Nomura: 0.86

Page 48
Part A.
1. <, >, <
2. =, <, >
3. <, <, =
4. >, =, <
5. >, <, >
Part B.

Race #1 Results	Race #2 Results
1st: Pablo	1st: Fran
2nd: Bill	2nd: Marina
3rd: James	3rd: Anna
4th: Jamal	4th: Jackie

Page 49
1. 9.6 cm
2. 3 cm 4 mm
3. 13 cm 8 mm
   13.8 cm
4. a) 22 mm or 2.2 cm
   b) 7 cm 6 mm or 7.6 cm

Page 50
1. $\frac{6}{8}$ or six-eighths
2. $\frac{3}{5}$ or three-fifths
3. $\frac{7}{10}$ or seven-tenths
4. $\frac{2}{4}$ or two-fourths
5. $\frac{2}{5}$ or two-fifths
6.

7.
8.
9. $\frac{60}{100}$ or sixty-hundredths

Page 51
1. $\frac{1}{2} = \frac{3}{6}$
2. $\frac{1}{3} = \frac{2}{6}$
3. $\frac{1}{2} = \frac{4}{8}$
4. $\frac{1}{2} = \frac{2}{4}$
5. $\frac{4}{8} = \frac{2}{4}$
6. $\frac{5}{8} = \frac{10}{16}$

Page 52
1. $\frac{1}{3}, \frac{1}{4}, \frac{2}{3}, \frac{2}{3}, \frac{3}{4}$
2. $\frac{2}{5}, \frac{2}{3}, \frac{3}{5}, \frac{5}{7}, \frac{3}{4}$
3. $\frac{1}{3}, \frac{7}{8}, \frac{1}{4}, \frac{5}{12}, \frac{7}{16}$

Page 53
1. a) $\frac{1}{3}$; b) $\frac{2}{3}$; c) $\frac{3}{4}$
2. a) $\frac{1}{5}$; b) $\frac{1}{2}$; c) $\frac{3}{5}$
3. a) $\frac{3}{8}$; b) $\frac{5}{8}$; c) $\frac{3}{4}$
4. a) $\frac{1}{4}$; b) $\frac{1}{3}$; c) $\frac{3}{4}$
5. a) $\frac{1}{13}$; b) $\frac{5}{13}$; c) $\frac{8}{13}$
6. a) $\frac{1}{2}$; b) $\frac{3}{4}$; c) $\frac{1}{1} = 1$
7. $\frac{2}{7}$
8. $\frac{3}{16}$

Page 54
1. $\frac{4}{8}; \frac{2}{6}; \frac{2}{8}; \frac{6}{9}; \frac{9}{12}$
2. $\frac{6}{15}; \frac{6}{8}; \frac{8}{12}; \frac{10}{14}; \frac{6}{18}$
3. $\frac{10}{15}, \frac{12}{30}, \frac{10}{28}, \frac{33}{36}, \frac{16}{28}$
4. $\frac{30}{42}, \frac{20}{24}, \frac{21}{24}, \frac{27}{36}, \frac{15}{24}$

Page 55
1. >, <, >, <
2. >, =, >, <
3. <, <, <, =

Pages 56–57
1. 2: 2, 4, 6, 8, 10
   3: 3, 6, 9, 12, 15
   4: 4, 8, 12, 16, 20
   5: 5, 10, 15, 20, 25
2. 6: 6, 12, 18, 24, 30
   8: 8, 16, 24, 32, 40
   10: 10, 20, 30, 40, 50
   12: 12, 24, 36, 48, 60
3. 12; 15; 6; 24
4. =, <, >, <
5. <, >, >, <
6. $\frac{7}{12}, \frac{2}{3}, \frac{3}{4}; \frac{2}{3}, \frac{13}{18}, \frac{7}{9}; \frac{7}{24}, \frac{1}{3}, \frac{3}{8}; \frac{6}{14}, \frac{1}{2}, \frac{4}{7}$

Page 58
1. $1\frac{1}{2}$ or $\frac{3}{2}$
2. $\frac{5}{4}$ or $1\frac{1}{4}$
3. $2\frac{3}{4}$ or $\frac{11}{4}$
4. $1\frac{3}{4}$ or $\frac{7}{4}$
5. $\frac{14}{8}$ or $1\frac{3}{4}$
6. $\frac{7}{3}$ or $2\frac{1}{3}$
7. improper fraction: $\frac{9}{4}$
   mixed number: $2\frac{1}{4}$
8. improper fraction: $\frac{19}{8}$
   mixed number: $2\frac{3}{8}$

Page 59
1. a) $\frac{2}{16}$ b) $\frac{4}{16}$ c) $\frac{8}{16}$ d) $\frac{16}{16}$
2. a) $\frac{1}{2}$ b) $\frac{3}{4}$ c) $\frac{9}{16}$ d) $\frac{3}{8}$
3. a) $4\frac{10}{16} = 4\frac{5}{8}$ b) $4\frac{5}{8}$ c) $4\frac{2}{4} = 4\frac{1}{2}$ or $4\frac{3}{4}$ d) $4\frac{1}{2}$

Page 61
1. 11% shaded, 89% unshaded, 100% total
2. 30% shaded, 70% unshaded, 100% total
3. 53% shaded, 47% unshaded, 100% total
4. 47% shaded, 53% unshaded, 100% total
5. a) $40,000
   b) 85%

6. a) 58%
   b) 75%
7. a) 300,000
   b) 300%
8. a) $2,000,000
   b) 400%

Pages 62–63

1. 0.25; 0.37; 0.5; 0.78; 0.98
2. 3; 6; 2; 4; 5
3. $\frac{1}{5}, \frac{1}{4}, \frac{1}{2}, \frac{3}{5}, \frac{3}{4}$
4. d: 0.2; f: $\frac{1}{5}$
5. d: 0.05; f: $\frac{1}{20}$
6. d: 0.12; f: $\frac{3}{25}$
7. T; F; F
8. F; T; T
9. a) 60%
   b) $\frac{2}{5}$
10. a) 60%
    b) 0.6
11. a) $\frac{9}{10}$
    b) 30%
    c) 0.63
12. a) $\frac{3}{4}$
    b) 75%
13. a) 0.25
    b) $\frac{1}{4}$
14. a) 0.45
    b) $\frac{45}{100}$
    c) 45%

Page 65

1. $0.32
2. $0.20
3. yes
4. Flats: $\frac{2}{12}$; Pumps: $\frac{3}{12}$; Sandals: $\frac{2}{12}$; Thongs: $\frac{1}{12}$; Tennis Shoes: $\frac{4}{12}$
5. Tennis Shoes
6. $\frac{2}{12}$ or $\frac{1}{6}$
7. 65%
8. 15%
9. 16%

Pages 66-67

1. (2) two-thousandths
2. (3) 0.32
3. (4) 2.5 < 1.99
4. (1) Bill, Jess, Bobby, Al
5. (3) 1 and 4
6. (4) $\frac{2}{3}$
7. (4) 12
8. (2) A, B, C
9. (3) $\frac{8}{12}$

10. (3) $1\frac{11}{16}$
11. (2) 35%
12. (1) 25%
13. (4) 60%

Unit 4: Fractions

Pages 69–71

1. $1\frac{1}{2}$; 2; $2\frac{1}{2}$; $2\frac{1}{2}$; 2
2. 2 inches
3. $3\frac{1}{2}$
4. 17; 11; 26; 43; 54
5. 3; 7; 16; 6; 4
6. 72; 25; 140; 168
7. 4; 3; 8; 7
8. 9 pounds
9. $12
10. 14 inches
11. (2) $33 - (14 + 10)$
12. 24
13. 17
14. 9
15. (5) $\frac{15-2}{50}$ lb.

Page 72

1. $2\frac{1}{3}$; $1\frac{3}{10}$; $4\frac{1}{2}$; $3\frac{2}{3}$; $2\frac{4}{5}$
2. $1\frac{1}{2}$; $4\frac{1}{2}$; $3\frac{1}{2}$; $4\frac{3}{5}$; $4\frac{1}{3}$
3. Improper fraction: $\frac{42}{12} = \frac{21}{6} = \frac{7}{2}$
   Mixed number: $3\frac{1}{2}$
4. Improper fraction: $\frac{18}{8} = \frac{9}{4}$
   Mixed number: $2\frac{1}{4}$

Pages 73–74

1. $\frac{2}{3}$; $\frac{3}{4}$; $\frac{7}{8}$; $\frac{4}{5}$; $\frac{15}{16}$
2. $\frac{1}{2}$; $\frac{1}{2}$; $\frac{3}{4}$; $\frac{1}{2}$; $\frac{2}{3}$
3. $\frac{3}{4}$; $\frac{8}{9}$; $\frac{4}{5}$; $\frac{2}{3}$; $\frac{5}{6}$
4. 1; 1; 1; 1; 1
5. 2; 2; 3
6. $1\frac{1}{4}$; $1\frac{1}{3}$; $1\frac{5}{8}$; $1\frac{5}{12}$; $1\frac{1}{6}$
7. $1\frac{1}{2}$; $1\frac{1}{4}$; $1\frac{2}{3}$; $1\frac{1}{2}$; $1\frac{1}{6}$
8. $2\frac{1}{3}$; $3\frac{1}{3}$; $2\frac{1}{4}$; 6; $3\frac{1}{4}$
9. $6\frac{1}{2}$; $4\frac{1}{2}$; 7; $9\frac{2}{3}$; $15\frac{7}{8}$

Page 75

1. $\frac{1}{4}$; $\frac{3}{8}$; $\frac{2}{5}$; $\frac{2}{9}$; $\frac{3}{10}$
2. $\frac{1}{2}$; $\frac{2}{3}$; $\frac{1}{2}$; $\frac{1}{4}$
3. $2\frac{1}{2}$; $1\frac{1}{2}$; $3\frac{1}{3}$; $1\frac{2}{5}$; $2\frac{2}{3}$
4. $1\frac{1}{2}$; $2\frac{3}{4}$; $3\frac{1}{3}$; $8\frac{5}{8}$; $8\frac{1}{6}$

Page 76

1. $\frac{3}{3}$; $\frac{4}{4}$; $\frac{12}{12}$; $\frac{2}{2}$; $\frac{6}{6}$
2. $2\frac{8}{8}$; $1\frac{3}{3}$; $4\frac{6}{6}$; $6\frac{2}{2}$; $11\frac{4}{4}$
3. $\frac{1}{4}$; $\frac{3}{8}$; $\frac{9}{16}$; $\frac{2}{3}$; $\frac{1}{2}$
4. $1\frac{2}{3}$; $4\frac{1}{4}$; $2\frac{3}{8}$; $6\frac{7}{10}$; $5\frac{11}{16}$
5. $3\frac{2}{5}$; $8\frac{1}{3}$; $7\frac{1}{8}$; $10\frac{3}{10}$; $11\frac{5}{16}$

Page 77

1. $2\frac{1}{2}$; $1\frac{2}{3}$; $\frac{4}{5}$
2. $2\frac{1}{2}$; $1\frac{1}{3}$; $1\frac{3}{8}$

3. $\frac{1}{3}$; $3\frac{5}{8}$; $4\frac{1}{2}$

4. $5\frac{3}{4}$; $8\frac{7}{8}$; $7\frac{3}{4}$

Pages 78–79

1. $\frac{3}{4}$; $\frac{5}{6}$; $1\frac{3}{8}$; $1\frac{3}{8}$; $1\frac{9}{16}$
2. $\frac{3}{10}$; $\frac{1}{8}$; $\frac{1}{2}$; $\frac{1}{8}$; $\frac{1}{16}$
3. $\frac{11}{12}$; $\frac{1}{15}$; $1\frac{4}{15}$; $\frac{1}{12}$; $\frac{1}{14}$
4. $\frac{1}{12}$; $\frac{5}{6}$; $\frac{3}{10}$; $1\frac{11}{20}$; $1\frac{7}{12}$
5. $5\frac{1}{2}$; $6\frac{3}{4}$; $10\frac{5}{8}$; $6\frac{3}{4}$
6. $8\frac{1}{2}$; $13\frac{1}{4}$; $7\frac{1}{2}$; $16\frac{9}{16}$
7. $12\frac{5}{12}$; $13\frac{1}{6}$; $16\frac{4}{15}$; $26\frac{7}{20}$
8. $4\frac{1}{8}$; $3\frac{1}{2}$; $3\frac{2}{3}$; $4\frac{1}{4}$
9. $2\frac{5}{8}$; $1\frac{1}{2}$; $4\frac{2}{3}$; $4\frac{3}{4}$
10. $4\frac{7}{12}$; $\frac{5}{6}$; $7\frac{7}{10}$; $8\frac{8}{15}$

Pages 80–81

1. $10\frac{3}{4}$ miles
2. $21\frac{13}{16}$ inches
3. $5\frac{1}{3}$ yards
4. $1\frac{3}{8}$ inches
5. $\frac{1}{8}$ ton
6. $3\frac{11}{12}$ pounds
7. $1\frac{5}{6}$ pounds
8. 7 inches
9. $1\frac{3}{8}$ cups
10. $4\frac{13}{16}$ inches
11. $4\frac{5}{12}$
12. a) length: $21\frac{7}{8}$ inches
    width: $11\frac{1}{4}$ inches
    b) $10\frac{5}{8}$ inches
13. Blakely: $4\frac{2}{3}$ hours
    Craig: $2\frac{11}{12}$ hours
14. $\frac{1}{15}$
15. a) $5\frac{1}{5}$ miles
    b) $6\frac{3}{5}$ miles
16. (4) $y = (\frac{5}{2} + \frac{5}{4} + \frac{7}{8}) \div 3$

Page 83

1. the number of calories in a serving of french fries
2. the percent of the class that's made up of 5-year-olds, *or* the percent that's made up of 6-year-olds
3. the fraction of Dave's salary that goes for utilities
4. the price that sirloin steak was reduced *to*
5. the number of monthly payments he must make
6. the weight of the bag of pears
7. the amount of time Lucas studies math each Sunday, Tuesday, and Thursday

Pages 84–85

1. $\frac{1}{3}$; $\frac{1}{4}$; $\frac{7}{15}$
2. $\frac{1}{15}$; $\frac{3}{16}$; $\frac{7}{20}$
3. $\frac{1}{2}$; $\frac{9}{70}$; $\frac{5}{8}$
4. $\frac{9}{16}$; $\frac{3}{20}$; $\frac{11}{16}$
5. $\frac{1}{10}$; $\frac{1}{4}$; $\frac{1}{10}$
6. $\frac{5}{8}$; $\frac{11}{30}$; $\frac{3}{4}$
7. $\frac{2}{3}$; $\frac{1}{10}$; $\frac{1}{6}$
8. $\frac{7}{36}$; $\frac{1}{5}$; $\frac{5}{8}$
9. $\frac{1}{4}$; $\frac{1}{3}$; $\frac{1}{2}$

Pages 86–87

1. $\frac{2}{1}$; $\frac{4}{1}$; $\frac{7}{1}$; $\frac{5}{1}$; $\frac{3}{1}$; $\frac{12}{1}$
2. $\frac{5}{3}$; $\frac{13}{4}$; $\frac{43}{8}$; $\frac{23}{5}$; $\frac{25}{3}$; $\frac{23}{2}$
3. 2; $3\frac{1}{3}$; $2\frac{2}{5}$
4. 6; 3; 8
5. 8; 16; 15
6. $\frac{11}{2}$; $\frac{23}{8}$; $\frac{23}{5}$
7. $10\frac{2}{3}$; $5\frac{1}{4}$; 27
8. $19\frac{1}{2}$; $13\frac{1}{4}$; 64
9. $1\frac{2}{3}$; $1\frac{3}{10}$; $2\frac{11}{32}$
10. $2\frac{1}{32}$; $3\frac{3}{4}$; $1\frac{8}{15}$
11. $9\frac{1}{3}$; $7\frac{7}{16}$; $19\frac{15}{16}$
12. $16\frac{19}{24}$; $4\frac{23}{64}$; $14\frac{7}{16}$

Pages 88–89

1. $\frac{3}{2}$; $\frac{8}{5}$; $\frac{6}{5}$; $\frac{2}{1}$; $\frac{4}{3}$
2. $\frac{3}{4}$; $\frac{5}{7}$; $\frac{4}{9}$; $\frac{2}{5}$; $\frac{16}{21}$
3. $\frac{1}{2}$; $\frac{1}{3}$; $\frac{1}{5}$; $\frac{1}{4}$; $\frac{1}{7}$
4. $\frac{3}{8}$; $\frac{4}{13}$; $\frac{3}{13}$; $\frac{8}{13}$; $\frac{16}{35}$
5. $\frac{2}{3}$; $2\frac{1}{4}$; $\frac{8}{9}$; $1\frac{1}{3}$
6. 1; $1\frac{1}{2}$; $2\frac{3}{4}$; 7
7. $1\frac{1}{3}$; $1\frac{3}{4}$; 1; 5
8. $2\frac{1}{4}$; $4\frac{2}{3}$; $\frac{3}{5}$; $\frac{8}{9}$

Pages 90–91

1. $\frac{5}{16}$; $\frac{3}{16}$; $\frac{2}{15}$; $\frac{1}{8}$
2. $1\frac{3}{8}$; $\frac{5}{8}$; $\frac{1}{2}$; $\frac{2}{9}$
3. 8; 9; 6; $18\frac{2}{3}$
4. 12; 12; $10\frac{2}{3}$; $22\frac{1}{2}$
5. $1\frac{1}{2}$; $\frac{3}{7}$; 2; $\frac{2}{9}$
6. $2\frac{4}{5}$; $4\frac{1}{8}$; $\frac{6}{11}$; $\frac{6}{35}$
7. $1\frac{1}{6}$; $\frac{7}{16}$; $1\frac{5}{16}$; $\frac{14}{15}$
8. $1\frac{2}{3}$; $\frac{1}{2}$; $\frac{2}{3}$; $2\frac{16}{19}$
9. $\frac{7}{16}$; $1\frac{3}{8}$; $\frac{2}{3}$; $2\frac{6}{25}$

Pages 92–93

1. $\frac{7}{15}$ mile
2. 10
3. $2\frac{5}{6}$ hours
4. 5
5. 50
6. 18
7. 9 (with part of the 9th strip left over)
8. $1\frac{1}{4}$ feet
9. $34\frac{1}{4}$ inches
10. (5) $(\frac{5}{4})(\$3) + (\frac{13}{2})(\$4)$
11. Rent: $\frac{1}{4}$
    Food: $\frac{1}{5}$
    Auto: $\frac{3}{20}$
12. $64

13. $\frac{17}{24}$

14. Earrings: 9
    Bracelets: 5
    Buckles: 6

Pages 94–95

1. $\frac{2}{3}$; $\frac{3}{4}$; $\frac{5}{1}$; $\frac{3}{2}$; $\frac{1}{2}$
2. $\frac{3}{1}$; $\frac{3}{2}$; $\frac{3}{2}$; $\frac{11}{6}$; $\frac{3}{2}$
3. a) $\frac{2}{3}$
   b) $\frac{3}{2}$
4. a) $\frac{4}{11}$
   b) $\frac{4}{7}$
5. $\frac{3}{40}$

6. a) $\frac{7}{8}$
   b) $\frac{8}{7}$
   c) $\frac{8}{15}$
7. a) $\frac{3}{1}$
   b) $\frac{1}{9}$
8. a) $\frac{7}{16}$
   b) $\frac{7}{9}$

Page 97

1. 76 beats per minute
2. 190 miles per hour
3. $33 per day
4. 16 miles per hour
5. #1: 24 miles per gallon
   #2: 26 miles per gallon
   #3: 20 miles per gallon
   #4: 18 miles per gallon
6. $\frac{5}{18}$ inch per hour
7. $11.50 per hour
8. $3\frac{1}{2}$ miles per hour
9. 4,500 gallons per hour
10. a) 7¢ per ounce
    b) 6¢ per ounce
    c) 6¢ per ounce
    d) 5¢ per ounce

Page 99

1. 19 (link: $+ 3\frac{1}{2}$)
2. (2) 5, 7, 11, 19, . . .
3. 162 (link: $\times$ 3)
4. (4) add 6, subtract 3
5. after 3 years (salary = $17,250)
6. after 4 weeks (height = $9\frac{3}{4}$ inches)

Page 101

1. Pro Sport
2. 450
3. 225
4. $147,600
5. Road Classic and Mountain Ranger
6. 2,850
7. 275
8. $\frac{6}{7}$
9. (4) the cost of making each Pro Sport and each All Terrain model

Pages 102–103

1. (3) 23 in.
2. (4) $13 + 18 \times 2$
3. (2) $3\frac{1}{2}$
4. (1) $1\frac{5}{8}$
5. (4) 110
6. (3) 5
7. (3) 8
8. (5) 5 to 2
9. (1) 400 students per hour
10. (4) $21\frac{1}{4}$ in.
11. (3) 12:10 P.M.
12. (1) 3 to 2

Unit 5: Decimals

Page 105

1. a) 4 meters
   b) $10
   c) 7 centimeters
2. a) $24
   b) 6 ounces
   c) 29 miles per gallon

In 3–4, any reasonable estimate is acceptable. Sample estimates are given here.

3. $11; 2; $180
4. 10; 8; 14
5. 16.542
6. 60.165
7. 15.085
8. 8.12
9. (4) 23.375
10. (2) $287.66
11. (2) $6.26
12. (1) 6

Page 107

1. 0.5; 0.5; 0.9; 0.2; 0.9
2. 2.4; 1.8; 3.9; 5.1; 6.8
3. 0.35; 0.74; 0.28; 0.89; 0.35
4. 3.63; 4.05; 8.50; 6.91; 12.66
5. 0.354; 0.463; 2.938; 3.142; 27.006
6. a) $10.83
   b) $7.09
   c) $11.14
   d) $7.73
   e) $25.20
   f) $28.18

Page 108

1. 0.9; 1.47; 4.1; $1.05; $4.67
2. 1.465; 4.71; 11.505; 18.025; 10.475
3. 19.4; 4.1; 13.6

Page 109

1. 0.5; 0.16; 1.8; 8; $8.99
2. 0.525; 5.85; 2.285; 13.317; 13.265
3. $0.35 \approx 0.4$; $5.15 \approx 5.2$; $8.45 \approx 8.5$; $9.625 \approx 9.6$
4. $0.6875 \approx 0.69$; $1.525 \approx 1.53$; $2.375 \approx 2.38$; $14.5625 \approx 14.56$

Pages 110–111

1. 5.25 meters
2. Yes (0.414 > 0.300)
3. 2.86
4. 3.65 miles
5. 217.1 centimeters
6. 1.125 inches
7. 4.3°F
8. 28.8
9. 0.625 second
10. 2.9 cm (2 cm 9 mm or 29 mm)
11. $3.36 million
12. 4.075 miles
13. (4) $p = 25 - (7.5 + 6.25 + 1.75)$
14. $5.90
15. 0.15 lb.
16. 0.1 lb.
17. after 4 years (7.4 million)

Pages 112–113

1. 7.6 miles
2. 7,000 feet
3. 17
4. 62°F
5. 1.1 miles
6. 130 miles

Pages 114–115

1. 7.15; 2.67; 3.40; 1.110
2. 22.5; 1.38; 10.5; $3.60; $29.61
3. 56.16; 35.7; 1.4288; $22.44; 6.825
4. 19.35 ≈ 19.4; 5.04 ≈ 5.0; 1.4878 ≈ 1.5; 15.75 ≈ 15.8
5. $49.248 ≈ $49.25; 166.805 ≈ 166.81; 18.750 = 18.75; $149.086 ≈ $149.09
6. 0.016; 2.08; 0.3312; 0.02412; 0.13833
7. 0.12; 0.054; 0.0216; 0.004

Pages 116–118

1. 2.08; 0.151; 8.2; 0.057; 0.206
2. 0.074; 0.053; 0.129; 0.085; 0.0021
3. $2.29; $2.75; $2.55; $0.04; $0.03
4. 23; 3; 149.7; 17.2; 5.2
5. 50; 2,200; 12; 1,100; 13
6. 300; 600; 200; 20; 155
7. 2.5; 4.8; 8.25; 4.75; 3.375
8. 0.5; 0.75; 0.7; 0.625; 0.4375
9. 0.33; 0.67; 0.17; 0.83; 0.56

Page 119

1. 5; 8.6; 37
2. 75; 260; 1,520
3. 800; 9,250; 25,600
4. 0.07; 0.65; 1.27
5. 0.287; 0.034; 0.005
6. 0.125; 0.067; 0.0098

Pages 120–121

1. (3) $7.80
2. 20.8 ≈ 21 miles per gallon
3. 109.5 ≈ 110 pounds
4. 19 miles
5. (5) $250 ÷ 20
6. $11.90
7. 0.7885 ≈ 0.79 ton
8. 0.0625 inch
9. (1) (24.5 − 1.6) ÷ 24
10. $18.40
11. #2: $\frac{11}{32}$ inch
12. 17.25 inches

13. 0.0505 (≈ 0.05 pound)
14. $286.80
15. (3) $\frac{8 - 3.28}{6}$

Page 123

1. a) 64
   b) 76
2. a) by 9:10 A.M.
   b) 7:30 A.M.
3. a) 3 pounds
   b) 6 pounds
4. 39°F
5. $3.45

Page 125

1. 1.7 liters per minute
2. 1.75 to 0.3 or about 6 to 1 (1.8 ÷ 0.3 = 6)
3. 0.45 m or 45 cm
4. about 20 cm
5. between her 3rd and 4th birthdays
6. the 1st year

Pages 126–127

1. (3) $20 − ($4 × 3)
2. (5) 36.5
3. (1) 13 cm
4. (2) 0.3125"
5. (4) 248.5
6. (3) 0.313
7. (2) $2.15
8. (3) $\frac{38" - (3 \times 0.5")}{4}$
9. (4) $\frac{1}{3}$
10. (2) 77 feet
11. (1) 3($6) + 4($12)
12. (3) 1:10 P.M.
13. (5) 88
14. (3) 4.6666666
15. (3) 0°C
16. (5) 15

Unit 6: Percents

Page 128

1. a) 20%
   b) 90
   c) 18
2. a) 60%
   b) $8.00
   c) $4.80
3. (2) whole
4. (1) percent

Page 129

1. a) 42
   b) 10%
   c) 420
2. a) 130
   b) 65%
   c) 200
3. a) $36
   b) 20%
   c) $180
4. P
5. %
6. %

Page 131

1. multiply the percent by the whole
2. divide the part by the whole
3. divide the part by the percent
4. a) P
   b) multiplication
5. a) %
   b) division
6. a) W
   b) division

7. a) %
   b) division
8. a) P
   b) multiplication

Page 133

1. $\frac{3}{10}, \frac{1}{2}, \frac{17}{20}, \frac{2}{3}, \frac{1}{3}$
2. $\frac{1}{100}, \frac{1}{20}, \frac{9}{100}, \frac{2}{25}, \frac{3}{50}$
3. 0.25; 0.5; 0.75; 0.9; 0.07
4. 0.055; 0.099; 0.085; 0.006; 0.001
5. a) $\frac{1}{4}$
   b) $\frac{1}{3}$
   c) $\frac{1}{10}$
   d) $\frac{3}{5}$
   e) $\frac{2}{3}$
   f) $\frac{3}{4}$

6. a) 5¢
   b) 7¢
   c) 12¢
   d) 25¢
   e) 50¢
   f) 75¢
   g) 90¢

Pages 134–135

1. 17; 30; 64; 15
2. 4.5; $84; 28.5; $2.75
3. $5.95; $13; 240; $9,500
4. $1.92
5. $7.50
6. a) $26.70
   b) $916.70

7. a) $17.60
   b) $337.60
8. 1,000
9. $8 million
10. $450
11. $375

Pages 136–137

1. 20%
2. 25%
3. 75%
4. $33\frac{1}{3}$%
5. 37.5% *or* $37\frac{1}{2}$%
6. 50%
7. 75%
8. 6%
9. 15%
10. $66\frac{2}{3}$%
11. a) $5,800
    b) 40%

12. 14%
13. (3) $\frac{92-26}{92} \times 100\%$
14. a) $16.00
    b) $15.00
    c) $16.00
    d) $66.00
15. a) 20%
    b) 25%
    c) $33\frac{1}{3}$%
    d) $66\frac{2}{3}$%

Pages 138–139

1. 300
2. $180
3. 90 tons
4. 225
5. 70
6. $1,100
7. $350
8. $300

9. a) 75%
   b) $36
10. a) 60%
    b) 150
11. a) 30%
    b) (1) $\frac{\$20}{0.3}$
12. a) 16,000
    b) 1,280
    c) $\frac{8}{5}$

Page 141

1. $1,316.10
2. $90 ($89.87)
3. $171.55
4. 1,900 calories per day

5. $17.55
6. $79.65
7. $231.20
8. $8,300

Page 143

1. $300
2. a) $72
   b) $872
3. $10,520
4. a) $346
   b) $8,304
   c) $804

5. a) $131
   b) $1,288
   c) $312 ($1,288 − $976)

Pages 144–145

1. $\frac{1}{3}$yr.; $\frac{1}{2}$yr.; $\frac{3}{4}$yr.; $\frac{11}{12}$ yr.
2. 0.25; 0.33; 0.75; 0.83
3. $\frac{15}{12} = \frac{5}{4}$ yr.; ≈ 2.33 yr.; $\frac{33}{12}$yr. = $\frac{11}{4}$yr.
4. $15
5. $43.33
6. $54
7. $1,062.50
8. $1,867.50
9. $805
10. a) May: $1.20
    June: $1.80
    July: $3.00
    August: $5.25
    b) $72.00
11. Diamond Visa at $8 is less expensive than National Visa at $9.

Page 147

Any reasonable estimate is acceptable. Typical answers are given.

1. ≈ $300 (25% of $1,200 *or* $1,200 ÷ 4)
2. ≈ $500 ($50 ÷ 10% *or* $50 × $\frac{10}{1}$)
3. ≈ 20% ($0.70 ÷ $3.50)
4. ≈ 300 ($33\frac{1}{3}$% × 900 *or* 900 ÷ 3)
5. (2) a little more than 35¢ per hour
6. ≈ 200 (600 ÷ 3)
7. ≈ $1,400 ($350 ÷ 25% or $350 × $\frac{4}{1}$)
8. a) ≈ $120 (25% ≈ $40, so 75% ≈ $120)
   b) ≈ $1,440 (12 × $120)
9. (4) ($15 ÷ 10) + $\frac{1}{2}$($15 ÷ 10)

Page 149

1. 25%; 40%; 37.5%; 70%; $33\frac{1}{3}$%; $83\frac{1}{3}$%
2. 50%; 30%; 25%; 45%; 87.5%
3. 27.5%; 6%; 175%; 450%; 600%
4.

Percent	Decimal	Fraction	Percent	Decimal	Fraction
10%	0.1	$\frac{1}{10}$	60%	0.6	$\frac{3}{5}$
20%	0.2	$\frac{1}{5}$	$66\frac{2}{3}$%	$0.66\frac{2}{3}$	$\frac{2}{3}$
25%	0.25	$\frac{1}{4}$	70%	0.7	$\frac{7}{10}$
30%	0.3	$\frac{3}{10}$	75%	0.75	$\frac{3}{4}$
$33\frac{1}{3}$%	$0.33\frac{1}{3}$	$\frac{1}{3}$	80%	0.8	$\frac{4}{5}$
40%	0.4	$\frac{2}{5}$	90%	0.9	$\frac{9}{10}$
50%	0.5	$\frac{1}{2}$	100%	1	$\frac{1}{1}, \frac{2}{2}, \frac{3}{3}$, etc.

5. $33\frac{1}{3}\%$
6. a) 85%
   b) 15%
7. 4.5%
8. a) 62.5%
   b) 37.5%

Page 151

1. pork
2. tuna, eggs, and lamb
3. a) ≈ 10%
   b) ≈ 0.4 ounce
4. ≈ 90%
5. lettuce and watermelon
6. 14 ounces
7. 3 ounces
8. $\frac{2}{1}$
9. ≈ $0.20

Pages 152–153

1. (1) 133 ÷ 0.35
2. (2) B and C only
3. (4) 194
4. (1) $25.20
5. (5) 16%
6. (3) $412.50
7. (4) $39
8. (2) $36 + ($\frac{1}{3}$ × $36)
9. (4) $\frac{\$290 - \$240}{\$290} \times 100\%$
10. (2) $9,660
11. (5) $18,750
12. (3) $3,000 + ($3,000 × $\frac{1}{10}$ × 2)
13. (3) 34,200
14. (5) 83,000

Unit 7: Data Analysis and Probability

Page 155

In 1–4, any reasonable estimate is acceptable. Typical estimates are given.

1. a) 155 pounds
   b) 152 pounds
2. a) 93°F
   b) 75°F
3. a) 20,000 pairs
   b) 42,500 pairs
4. a) 240 million
   b) 290 million

Pages 157–158

1. Answers are given in the table.

Millie's Famous Sandwich Spread (4-ounce serving)		
Ingredient	Ounces	Fat (g)
Chicken: Light Meat	1.8	1.8
Chicken: Dark Meat	1.2	2.4
Cream Cheese	0.8	8
Mayonnaise	0.2	2.2
Total Grams of Fat: 14.4		

2. Answers for part a and b are given in the table.

Activity	Oxygen Rate (Graph A)	Calorie Rate (Graph B)
Sitting	0.2	1
Walking	0.6	4
Bicycling	1.2	6
Swimming	1.6	7
Jogging	2.0	11

3. Answers are given in the following table.

Employee	Regular Pay (1st 40 hours)	Overtime Pay (Hours over 40)
Fenrick	$280	$42
Rodriquez	$280	$84
Smith	$280	$21
Yeh	$280	$63

4. a) ≈ $550 ($5\frac{1}{2}$ × 50 × $2)
   b) ≈ $500 (5 × 50 × $2)

Pages 160–161

1. (4)
2. (2) and (4)
3. (1) and (4)
4. (4)

Page 163

1. a) $\frac{1}{6}$
   b) $\frac{1}{2}$ ($\frac{3}{6}$)
   c) $\frac{5}{9}$ ($\frac{20}{36}$)
   d) 0
2. a) $\frac{1}{4}$ ($\frac{2}{8}$)
   b) $\frac{3}{8}$
3. a) $\frac{1}{9}$ ($\frac{4}{36}$)
   b) $\frac{1}{3}$ ($\frac{12}{36}$)
4. $\frac{7}{30}$
5. a) $\frac{1}{3}$
   b) $\frac{2}{3}$

Page 165

1. a) $\frac{5}{36}$
   b) $\frac{1}{18}$ $\left(\frac{2}{36}\right)$
2. a) The girls can be seated in 6 different ways:

#26	#27	#28
Amy	Gail	Vicki
Amy	Vicki	Gail
Gail	Amy	Vicki
Gail	Vicki	Amy
Vicki	Amy	Gail
Vicki	Gail	Amy

   b) $\frac{2}{3}$ $\left(\frac{4}{6}\right)$ From the list in part a, you can see that there are 4 seating arrangements in which Amy sits next to Vicki.
3. a) 9 combinations
   b) $\frac{1}{9}$
4. a) You are 2 times more likely to roll a 7 (6 ways) than a 4 (3 ways). $6 \div 3 = 2$
   b) $\frac{1}{6}$ $\left(\frac{6}{36}\right)$
5. a) The posters can be arranged in 6 different ways:

   ties   belts   shirts
   ties   shirts   belts
   belts   ties   shirts
   belts   shirts   ties
   shirts ties   belts
   shirts belts   ties

   b) $\frac{1}{3}$ $\left(\frac{2}{6}\right)$ From the list in part a, you can see that there are only 2 arrangements in which ties are *not* next to shirts.
6. a) 6: GNB, GBN, NGB, NBG, BGN, BNG
   b) $\frac{1}{3}$ $\left(\frac{2}{6}\right)$
   c) $\frac{2}{3}$ $\left(\frac{4}{6}\right)$

Page 167

1. $\frac{1}{36}$ $\left(\frac{1}{6} \times \frac{1}{6}\right)$
2. $\frac{1}{4}$ $\left(\frac{1}{2} \times \frac{1}{2}\right)$
3. a) $\frac{1}{4}$ $\left(\frac{1}{2} \times \frac{1}{2}\right)$
   b) $\frac{1}{8}$ $\left(\frac{1}{4} \times \frac{1}{2}\right)$ For part b, you consider the first two flips as a single event with a probability of $\frac{1}{4}$.
4. a) $\frac{2}{5}$
   b) $\frac{1}{10}$ $\left(\frac{2}{5} \times \frac{1}{4}\right)$
5. $\frac{1}{7}$ $\left(\frac{3}{7} \times \frac{2}{6}\right)$
6. $\frac{8}{33}$ If Audrey draws a blue marble first, the probability is $\frac{8}{12} \times \frac{4}{11} = \frac{8}{33}$. If Audrey draws a green marble first, the probability is $\frac{4}{12} \times \frac{8}{11} = \frac{8}{33}$.
7. $\frac{1}{6}$ $\left(\frac{5}{10} \times \frac{3}{9}\right.$ *or* $\left.\frac{3}{10} \times \frac{5}{9}\right)$
   You get the same answer whether you assume Bill draws a dime first *or* a nickel first.

8. a) $\frac{1}{5}$ $\left(\frac{3}{6} \times \frac{2}{5}\right)$
   b) $\frac{1}{20}$ $\left(\frac{1}{5} \times \frac{1}{4}\right)$ For part b, you consider the choosing of two heads-up pennies as a single event with the probability of $\frac{1}{5}$ (as determined in part a).

Page 169

1. about 83 $\left(\frac{1}{6} \times 500 = 83.3 \ldots\right)$
2. a) $\frac{9}{200}$
   b) about 34 $\left(\frac{9}{200} \times 750 = 33.75\right)$
3. a) $\frac{1}{4}$ $\left(\frac{250}{1000}\right)$
   b) 76 $\left(\frac{380}{1000} \times 200\right)$
4. a) 55% *or* $\frac{11}{20}$ $\left(\frac{55}{100}\right)$
   b) about 41 (55% of 75 is 41.25 or $\frac{11}{20} \times 75 = 41\frac{1}{4}$)
5. about 7 $\left(\frac{1}{36} \times 250 = 6\frac{34}{36}\right)$
6. a) $\frac{1}{10}$
   b) $\frac{1}{100}$ $\left(\frac{1}{10} \times \frac{1}{10}\right)$
7. a) $\frac{1}{10}$ $\left(\frac{4}{10} \times \frac{1}{4}\right)$
   b) 3 $\left(\frac{1}{10} \times 30\right)$
8. a) 4 (20% of 20 *or* $\frac{1}{5} \times 20$)
   b) $\frac{9}{100}$ $\left(\frac{3}{10} \times \frac{3}{10}\right)$

Pages 170–171

1. (2) $15,000
2. (3) $25,000
3. (4) $60,000
4. (5) VCRs are more popular than cable TV.
5. (1) The cost of watching movies on a VCR is less than watching them on cable TV.
6. (5) 90% or more
7. (3) $\frac{1}{3}$
8. (1) $\frac{1}{9}$
9. (5) 50%
10. (2) 25
11. (2) 25%
12. (1) $\frac{1}{25}$

Unit 8: Algebra

Page 173

1. no
2. no
3. yes
4. yes
5. $x = 8$, $y = 4$, $n = 3$, $m = 8$
6. $y = 6$, $z = 12$, $x = 18$, $n = 10$
7. $z = 2\frac{1}{2}$, $x = 1\frac{1}{3}$, $y = \frac{4}{5}$, $n = 1\frac{11}{12}$
8. $x = 15$, $y = 17$, $z = 10$, $n = 2$
9. $m = 20$, $x = 18$, $y = 17$, $p = 8$
10. $x = 3\frac{5}{8}$, $y = 2\frac{1}{3}$, $z = \frac{25}{8}$ or $3\frac{1}{8}$, $n = 5\frac{7}{8}$

Page 175

1. no
2. yes

3. no

4. $y = 7, z = 8, n = 9, z = 6$

5. $x = 1.75, y = 2.6, z = \frac{13}{6}$ or $2\frac{1}{6}$, $n = \frac{7}{10}$

6. $x = 36, y = 45, z = 35, n = 84$

7. $n = 4\frac{1}{2}, x = 13.5, y = 14\frac{2}{5}, p = 19.6$

8. $x = 35, y = 16, z = 16, n = 48$

Page 177

1. $x = 1, y = 4, n = 1, z = 9$

2. $m = 4, p = 9, y = 2, x = 2\frac{1}{2}$

3. $x = 15, y = 30, z = 6, n = 80$

4. $z = 22.5, m = 10, y = 30, p = 37\frac{1}{2}$

5. $x = 12, y = 27, z = 10, n = 15$

6. $x = 20, z = \frac{9}{4}$ or $2\frac{1}{4}, x = 6, y = 6$

Page 179

Equations may vary.

1. a) x = Kari's monthly salary, b) $\frac{1}{4}x = \$375$, c) $x = \$1,500$

2. a) p = original price of set, b) $p - \$65 = \$320$, c) $p = \$385$

3. a) r = regular hourly rate, b) $8r + \$15 = \$71$, c) $r = \$7$

4. a) d = amount of usual deposit, b) $6d + \$150 = \$900$, c) $d = \$125$

In 5–7, any letter can be used as the variable.

5. a) t = total enrollment, b) $t = 180 \div 60\%$, c) $t = 300$

6. a) l = length of uncut ribbon, b) $l = 8(12 + 2)$, c) $l = 112$ in.

7. a) t = total amount of punch, b) $\frac{1}{4}t + 3 = 7$, c) $t = 16$ quarts

Page 181

1. $5x, 3n, 10y, 10z$

2. $\frac{3}{4}x, \frac{3}{8}y, 1\frac{1}{6}n, 1\frac{1}{4}w$

3. $x = 4, y = 3, z = 3, m = 9$

4. $x = 2\frac{1}{4}, n = 4, z = 3\frac{7}{11}, x = 12$

5. $z = 8.5, x = 20, y = 4, x = 8.2$

6. $n = 2, z = 7, y = 1\frac{5}{7}, x = 4\frac{2}{15}$

7. $x = 14, q = 5, m = 3, n = 2\frac{1}{2}$

Pages 182–183

1. $x = 4, z = 22, n = 7$

2. $y = 15, x = 9, p = 3$

3. $n = 3, y = 2, x = 5$

4. $b = 2, y = 3, x = 13\frac{1}{3}$

5. $x = 10, w = 9, z = 0$

6. $n = 12\frac{1}{2}, x = 32, y = 112$

Page 185

1. $4n + 2 = 2n + 8, n = 3$

2. $\frac{3}{4}x + \$150 = 2x, x = \$120$

3. 44, 88, and 132

4. Jo's share is $3.85, Anna's is $4.50

5. (4) $\frac{n}{4} + 11 = 3n$

6. Chris earns $1,100; Jean earns $1,300

7. 12, 13, and 14

8. 600

9. skirt profit = $3.50, blouse profit = $7.00

10. (3) $x + \frac{5}{4}x = \$2,140$

Page 187

1. no, no, yes

2. yes, no, yes

3. a) $\frac{5}{6} = \frac{15}{18}$

b) $\frac{3}{2} = \frac{9}{6}$

4. $\frac{5}{15} = \frac{1}{3}, \frac{4}{3} = \frac{32}{24}, \frac{x}{9} = \frac{24}{29}, \frac{4}{5} = \frac{28}{y}$

5. $x = 3, y = 30, n = 2, x = 4$

6. $h = 6, x = 4, y = 9, x = 60$

7. $x = 20, y = 7, h = 3, x = 12$

8. 12, 5, 72, 48

Page 189

1. $2\frac{1}{4}$ quarts

2. $296

3. $2.86

4. 382.5 miles

5. (3) 40

6. 252 miles

7. 21

8. 63.5 cm

9. 15

10. (5) $\frac{2}{5} = \frac{x}{135}$

Pages 190–191

1. a) dependent variable is y; independent variable is x

b) dependent variable is c; independent variable is s

c) dependent variable is m; independent variable is n

2. $y = 12, a = 9, m = 12$

3. 3, 4, 5, 6

4. 0, 2, 4, 6

5. 5, 6.5, 8, 9.5

6. 9, 17, 25, 33

7. 6, $7\frac{1}{2}$, 9, $10\frac{1}{2}$

8. 1, 3, 5, 7

Page 193

1. A = (2,3); B = (−5,1); C = (−4,−5); D = (4,−3)

2.

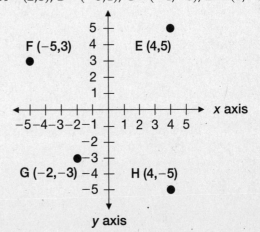

Pages 194–195

1. 0, 2, 4   2. 1, 2, 3

Graph for 1 and 2.

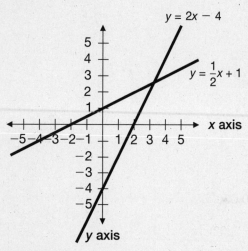

3. a) $11.25
   b) $2.50, $5.00, $7.50

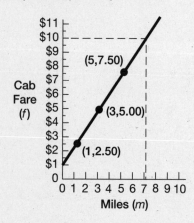

   c) a little more than 7 miles.
4. a) 212°F
   b) 59, 77, 95

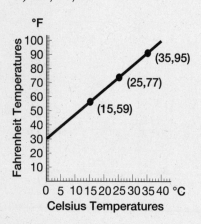

   c) 0°C

Page 197

1. Line A: positive
   Line B: zero

Line C: negative
Line D: undefined

2. slope = $-\frac{2}{4} = -\frac{1}{2}$
3. a) slope = $\frac{2}{2} = 1$
   b) x intercept is $(-2,0)$
   c) y intercept is $(0,2)$
4. b) slope = $\frac{4-0}{3-0} = \frac{4}{3}$ or $1\frac{1}{3}$
   c) slope = $\frac{8-5}{6-2} = \frac{3}{4}$

Pages 198–199

1. (4) Divide each side by 4.
2. (2) $0.75p = \$27$
3. (3) $x = 6$
4. (1) $4y - 2 = 2y + 6$
5. (3) 6
6. (2) $415
7. (3) $85
8. (4) $x + x + 2 + x + 4 = 42$
9. (1) 36
10. (3) 93
11. (5) $\frac{5}{7} = \frac{x}{98}$
12. (2) 1
13. (2) B
14. (5) $y = \frac{1}{4}x + 1$

Unit 9: Geometry

Page 200

1. ∠ RST or ∠ TSR
2. ∠ C
3. ∠ 4
4. ∠ XYZ or ∠ ZYX

Page 201

1. a) 60°
   b) ⟋ 60°
2. a) 36°
   b) ⟋ 36°
3. c

4. e
5. d
6. a
7. b

Page 203

1. acute
2. reflex
3. obtuse
4. straight
5. right
6. acute
7. 53°

8. 123°
9. 38°
10. 52°
11. 55°
12. 116°
13. 45°

Pages 204–205

1. 32°
2. 131°
3. 90°
4. ∠ b = 45°; ∠ c = 135°; ∠ d = 45°
5. ∠ 1 = 75°; ∠ 2 = 105°; ∠ 3 = 75°

6. $\angle x = 120^\circ$; $\angle y = 60^\circ$; $\angle z = 120^\circ$
7. $\angle A = 134^\circ$; $\angle B = 46^\circ$; $\angle C = 134^\circ$
8. 41°
9. F, H, I, N, Z
10. 360°

Page 207
1. isosceles
2. scalene (circled because it is also a right triangle)
3. equilateral
4. scalene
5. 79°
6. 132°
7. 114°
8. 104°
9. 43°
10. a) 51°
   b) 39° ($180^\circ - 90^\circ - 51^\circ$)
11. $\angle A = 144^\circ$; $\angle B = 36^\circ$
12. $\angle C = 30^\circ$; $\angle D = 60^\circ$; $\angle E = 90^\circ$

Page 209
1. $d = 20'$
2. $x = 20"$
3. 33 ft.
4. $d = 80$ ft.

Page 211
1. 12 in.
2. width: 156 inches or 13 feet
   length: 216 inches or 18 feet
3. a) height: 9 inches; width: 12 inches
   b) no
4. 3.2 inches ($3\frac{1}{5}$ inches)
5. about 1,240 miles (rounding 1,237.5 miles)
6. $3\frac{1}{4}$ inches (rounding 3.3 inches)

Page 213
1. 25
2. 64
3. 100
4. $\frac{1}{4}$
5. 12.25
6. $\frac{4}{9}$
7. 36
8. 10.24
9. $\frac{49}{64}$
10. 4
11. 10
12. 6
13. 13
14. 15
15. 14
16. 5
17. 7
18. 11
19. 61
20. 33
21. 5
22. 6
23. $\sqrt{50} \approx 7$
24. $\sqrt{35} \approx 6$
25. $\sqrt{90} \approx 9.5$
26. $\sqrt{12} \approx 3.5$
27. $\sqrt{98} \approx 10$
28. $\sqrt{155} \approx 12.5$

Page 215
1. 10 in.
2. 13 m
3. 12 yd.
4. 5 cm
5. 9 miles
6. 17 ft.
7. about 32 miles

Pages 216–217
1. 2^3; 2 to the 3rd power or 2 cubed; 8
2. 10^4; 10 to the 4th power; 10,000
3. 3^{-3}; 3 to the minus 3rd power; $\frac{1}{27}$
4. 10^{-4}; 10 to the minus 4th power; $\frac{1}{10,000}$ or 0.0001
5. 343
6. $\frac{1}{32}$
7. 1,000,000
8. $\frac{1}{1,000}$ or 0.001
9. 4×10^3
10. 8×10^6
11. 9×10^9
12. 2.5×10^4
13. 7.5×10^7
14. 1.38×10^6
15. 6×10^{-3}
16. 2×10^{-4}
17. 8×10^{-6}
18. 7.5×10^{-4}
19. 9.25×10^{-5}
20. 3.18×10^{-6}
21. 2.4×10^5 miles
22. 93,000,000 miles
23. 3.8×10^{-2} cm
24. 0.00000002 cm

Page 218
1. 32 miles
2. 210 cm
3. 16 in.

Page 219
1. 176 feet ($d = 56$ feet)
2. a) $24\frac{1}{2}$ or 24.5 inches
   b) about 77 inches
3. 20.4 meters
4. a) 6.6 yards
   b) about 21 yards

Pages 220–221
1. 81 m²
2. 70 sq. in.
3. 112 cm²
4. 56.25 sq. in.
5. 4 feet
6. a) $\frac{9}{16}$ sq. ft.
   b) 144 sq. ft.
   c) 256 tiles ($144 \div \frac{9}{16}$)
7. $16\frac{7}{8}$ sq. yd.
8. 50 feet
9. a) $\frac{4}{9}$ sq. ft.
   b) 96 sq. ft.
   c) 216 bricks ($96 \div \frac{4}{9}$)
10. 193 sq. ft. (49 + 144)
11. 37 sq. yd. (16 + 21)

Page 222
1. 22.5 sq. in.
2. 12.75 cm²
3. 7 sq. ft.
4. 90 sq. yd.
5. $18.00

Page 223
1. about 1,400 cm² (rounding 1,386)
2. about 19.6 sq. yd. (rounding 19.625)
3. about 7 sq. in. (rounding 7.065)
4. about 7,850 sq. mi.
5. 283 sq. ft.
6. $80.38
7. 9 to 1 or $\frac{9}{1}$

Page 224
1. 64 cu. yd.
2. 240 cu. ft.
3. 112.5 m³
4. a) 15 cu. ft.
   b) 930 pounds
5. 27 to 1 or $\frac{27}{1}$

Page 225

1. 22 cm^3
2. about 16,000 cu. in. (rounding 16,016)
3. about 175 cu. ft. (rounding 176.625)
4. a) 962 cu. ft.
   b) about 7,215 gallons
5. 9 to 1 or $\frac{9}{1}$

Pages 226–227

1. (2) 30°
2. (3) obtuse
3. (4) 180°
4. (2) 63°
5. (5) 71°
6. (2) 109°
7. (3) 65°
8. (1) 24 in.
9. (4) 550 miles
10. (5) 8.5
11. (2) 10 feet
12. (5) 132 feet
13. (3) 57%
14. (5) 22.5
15. (1) 38 cu. ft.